AF805754

Aviation Cybersecurity: Regulatory Approach in the European Union

Aviation Cybersecurity: Regulatory Approach in the European Union

Benjamyn I. Scott

eleven
international publishing

Published, sold and distributed by Eleven International Publishing
P.O. Box 85576
2508 CG The Hague
The Netherlands
Tel.: +31 70 33 070 33
Fax: +31 70 33 070 30
e-mail: sales@elevenpub.nl
www.elevenpub.com

Sold and distributed in USA and Canada
Independent Publishers Group
814 N. Franklin Street
Chicago, IL 60610
USA
Order Placement: (800) 888-4741
Fax: (312) 337-5985
orders@ipgbook.com
www.ipgbook.com

Eleven International Publishing is an imprint of Boom uitgevers Den Haag.

ISBN 978-94-6236-961-0
ISBN 978-94-6274-844-6 (E-book)

© 2019 The author | Eleven International Publishing

This publication is protected by international copyright law.
All rights reserved. No part of this publication may be reproduced, stored in a retrieval system, or transmitted in any form or by any means, electronic, mechanical, photocopying, recording or otherwise, without the prior permission of the publisher.

Table of Contents

List of Abbreviations

ACI	Airports Council International
AI	Artificial Intelligence
A-ISAC	Aviation Information Sharing and Analysis Center
AMC	Acceptable Means of Compliance
A-NPA	Advanced Notices of Proposed Amendment
ANS	Air Navigation Services
ANSP	Air Navigation Service Providers
APT	Advanced Persistent Threat
ASA	Air Service Agreement
ATC	Air Traffic Control
ATM	Air Traffic Management
AV-CERT	Aviation Computer Emergency Response Team
AVSEC	Aviation Security
AVSECP	Aviation Security Panel
BCN	Biological, Chemical and Nuclear
CAA	Civil Aviation Authority
CANSO	Civil Air Navigation Services Organization
CAT	Commercial Transport Aircraft
CERT	Computer Emergency Response Team
CERT-CC	Computer Emergency Response Team – Coordination Center
CERT-EU	Computer Emergency Response Team – European Union
CIA	Central Intelligence Agency
CMU	Carnegie Mellon University
CNS	Communication Navigation Surveillance
CS	Certification Specifications
CS-23	Certification Specification Normal, Utility, Aerobatic and Commuter Aeroplanes
CS-25	Certification Specification Large Aeroplanes
CS-27	Certification Specification Small Rotorcraft
CS-29	Certification Specification Large Rotorcraft
CS-E	Certification Specification Engines
CS-ETSO	Certification Specification European Technical Standard Orders
CS-P	Certification Specification Propellers
CSIRT	Computer Security Incident Response Teams
CTED	Counter-Terrorism Committee Executive Directorate
CUPPS	Common Use Passengers Processing System
DAA	Detect and Avoid Systems
DDoS	Disruptive Denial of Service
DG-MOVE	Directorate-General for Mobility and Transport
Doc.	Document
DoS	Denial of Service
EACCC	European Aviation Crisis Coordination Cell
EASA	European Aviation Safety Agency
EASTI	European Aviation Security Training Institute
EATM-CERT	European Air Traffic Management Computer Emergency Response Team
EC3	European Cybercrime Centre

ECAC	European Civil Aviation Conference
ECCSA	European Centre for Cybersecurity in Aviation
ECSCG	European Cyber Security for Aviation Standards Coordination Group
ED	Executive Director
EDA	European Defence Agency
e.g.	*Exempli Gratia* ('For Example')
ENISA	European Union Agency for Network and Information Security
EPAS	European Plan for Aviation Safety
ESCP	European Strategic Coordination Platform
EU	European Union
EU-LISA	European Agency for the Operational Management of Large-Scale IT Systems
EUROCAE	European Organisation for Civil Aviation Equipment
EVT	Evaluation Task
FAA	Federal Aviation Administration
FBI	Federal Bureau of Investigation
FMS	Flight Management System
FOT	Focused Attention Topic
FP9	Research Framework Programme
ft	Foot/Feet
GANP	Global Air Navigation Plan
GASeP	Global Aviation Security Plan
GASP	Global Aviation Safety Plan
GBAS	Ground Based Augmentation System
GBP	British Pound
GDP	Gross Domestic Product
GDPR	General Data Protection Regulation
GM	Guidance Material
GNSS	Global Navigation Satellite System
GPS	Global Positioning System
IATA	International Air Transport Association
ICAO	International Civil Aviation Organization
ICASS	ICAO Comprehensive Aviation Security Strategy
ICCAIA	International Coordinating Council of Aerospace Industries Associations
ICT	Information and Communication Technology
IFALPA	International Federation of Air Line Pilots
IFR	Instrument Flight Rules
IHLG	Industry High Level Group
IoT	Internet of Things
ISIL	Islamic State of Iraq and the Levant
ISS	International Space Station
IT	Information Technology
JAA	Joint Aviation Authorities
JAA LO	Joint Aviation Authorities Liaison Office
JAA T	Joint Aviation Authorities Transition
JAA TO	Joint Aviation Authorities Training Office
kg	Kilogramme(s)
m	Metre(s)
MBD	Management Board Decision
MoC	Memorandum of Cooperation
MoU	Memorandum of Understanding
MST	Member States Tasks

NAA	National Aviation Authority
NIS	Network and Information Systems
NPA	Notices of Proposed Amendment
NSA	National Security Agency
PANS	Procedures for Air Navigation Services
PASO	Pacific Aviation Safety Office
PPP	Public-Private Partnership
RES	Research Actions
RMT	Rulemaking Task
RPAS	Remotely Piloted Aircraft Systems
RSOO	Regional Safety Oversight Organisation
SAGAS	Stakeholder Advisory Group for Aviation Security
SARI	South Asia Regional Initiative
SARP	Standard and Recommended Practice
SASP	Separation and Airspace Safety Panel
SeMS	Security Management System
SERA	Standardised European Rules of the Air
SESAR	Single European Sky Air Traffic Management Research
SJU	SESAR Joint Undertaking
SOC	Security Operations Centre
SPT	Safety Promotion
STP	Standardised Training Packages
SUPP	Regional Supplementary Procedures
SWIM	System Wide Information Management
TCAS	Traffic Collision Avoidance System
TFEU	Treaty on the Functioning of the European Union
ToR	Terms of Reference
UAE	United Arab Emirates
UIC	Council and Committee on Unlawful Interference
UK	United Kingdom
UN	United Nations
UN CTC	UN Counter Terrorism Committee
US	United States
USD	United States Dollar
VFR	Visual Flight Rules
VLOS	Visual Line of Sight
VPN	Virtual Private Network
WiFi	Wireless Fidelity
WP	Working Paper

Acknowledgements

I wish to thank each and every person who has dedicated their free time, patiently dealt with my requests and provided fascinating information. In particular, I would like to pay special thanks to Cagri Aslan, Laura Hughes Gerber, Deepika Jey, Stefan Kaiser, Angeliki Karakoliou, Henning Koeppen, Ping-Yu Lim, Jean-Paul Moreaux, Mateusz Wojciech Osiecki and John Synnott for reviewing or discussing parts of this work.

My thanks also go out to the members of staff at the Institution of Air Law, Space Law and Cyber Law at the University of Cologne who have helped me at various points during this project with accessing literature, offering a forum for discussion and providing a welcome distraction from writing. Further, I would like to thank the European Aviation Safety Agency, Joint Aviation Authorities Training Organisation, Pacific Aviation Safety Office and South Asia Regional Initiative for kindly responding to my requests for information.

Finally, I would like to express my deep gratitude to my parents who have supported and encouraged me throughout.

Without these people, this work would never have come to fruition; thank you.

Introduction

The global aviation industry continues to grow. The International Air Transport Association (IATA), for example, announced that airlines' net profit is predicted to rise from USD 34.5 billion in 2017 to USD 38.4 billion in 2018.[1] IATA cited low oil prices, strong passenger demand, efficiency gains and reduced interest payments as major factors behind this positive financial forecast. Due to this, 2018 marked the fourth consecutive year of sustainable profits.[2] The trend is expected to continue as, for example, the predicted net profit is set to be USD 35.5 billion for 2019,[3] and other forecasts predict that 7.2 billion air travellers will fly in 2035, which is just under double the expected 4.3 billion for 2018.[4] One contributing factor towards high passenger demand is the public's perception of the industry as safe and secure. This is demonstrated by the significant reduction in passenger numbers on Malaysia Airlines flights in the wake of the MH370[5] and MH17[6] disasters. Malaysia Airlines' net loss grew to USD 137.8 million following the disappearance of flight MH370, which was partly due to reduced sales as, for example, ticket sales from China reduced by 60 percent in March 2014. Recognising the importance of this, the airline stated: "Winning back customers and a relentless cost focus will be part of the airlines' recovery plan."[7] Thus, if the public does not have confidence in the airline, or more broadly the industry, then this will have an adverse effect on demand and profitability.

---

1 IATA, 'Strong Airline Profitability Continues in 2018', 5 December 2017, http://www.iata.org/pressroom/pr/Pages/2017-12-05-01.aspx. The amount was later revised down from USD 33.8 billion to USD 32.3 billion in June 2018.

2 *Ibid.*

3 IATA, 'IATA Forecasts $35.5bn Net Profit for Airlines in 2019', https://airlines Aviation Stakeholder: Airlines.iata.org/news/iata-forecasts-355bn-net-profit-for-airlinesAviation Stakeholder:Airlines-in-2019.

4 IATA, 2017.

5 Malaysia Airlines Flight MH370 was a scheduled international passenger flight that disappeared on 8 March 2014 while flying from Kuala Lumpur, Malaysia, to Beijing, China. The Boeing 777-200ER aircraft, with 227 passengers and 12 crew on board, left the range of Malaysian military radar while over the Andaman Sea. While some debris from the aircraft has been found, the fate of MH370 has not been determined and the cause of its disappearance has yet to be established. It is feared that all 239 persons on board died with the total loss of the aircraft.

6 Malaysia Airlines Flight MH17 was a scheduled passenger flight from Amsterdam, The Netherlands to Kuala Lumpur, Malaysia. The Boeing 777-200ER was shot down by a Buk surface-to-air missile on 17 July 2014 while flying over eastern Ukraine. All 283 passengers and 15 crew members were tragically killed. *See*, for more information, Dutch Safety Board, MH17 crash report, 13 October 2015. *See*, Chapter III.3.2.

7 D. Kedmey, 'This Chart Shows What Happened to Malaysia Airlines After Missing Flight', Time, 15 May 2014, http://www.time.com/100823/malaysia-airlines Aviation Stakeholder:Airlines-decline/.

According to the Aviation Safety Network there were 14 fatal crashes in 2015 which resulted in 186 deaths.[8] This marked the lowest number of fatal crashes ever recorded in one year, which led 2015 to be branded the safest year ever for aviation. This list does not, however, take into consideration acts of sabotage or the shooting down of civilian aircraft. Germanwings Flight 9525[9] and Metrojet Flight 9268,[10] therefore, do not feature in this list.

Both the number of crashes per year and the total number of fatalities are generally declining. For example, 2016 was the second safest year according to Aviation Safety Network's statistics with 17 fatal losses of aircraft amounting to a total 258 deaths. However, 2015 was replaced as Aviation Safety Network data shows that 2017 was the safest year ever for commercial aviation. There was a total of 14 fatal airliner accidents, which resulted in 59 fatalities.[11] Unfortunately, 2018 did not surpass the 2017 record due to two crashes taking place in February.[12]

A significant contribution to the reduction in the number of fatal crashes and deaths is due to the elaborate safety regulations and oversight emanating from the International Civil Aviation Organization (ICAO), regional entities like the European Aviation Safety Agency (EASA or Agency) and national civil aviation authorities (CAAs). This includes, *inter alia*, extensive initial and continuing airworthiness testing, consistent rules of the air, sufficient personnel training, and aircraft accident and incident investigation to allow lessons to be learnt from past mistakes. However, as the regulatory aftermath of Germanwings Flight 9525 and the 9/11 terrorist attacks in the United States (US) show,

8 All accident and death figures are taken from the Aviation Safety Network's table on 'Accidents and Fatalities per Year'. Aviation Safety Network, 'ASN Data Show 2017 was Safest Year in Aviation History', 30 December 2017, https://www.news.aviation-safety.net/2017/12/30/preliminary-asn-data-show-2017-safest-year-aviation-history/.

9 Germanwings Flight 9525 was a scheduled international passenger flight from Barcelona, Spain to Düsseldorf, Germany which crashed on 24 March 2015 into the French Alps. All 144 passengers and 6 crew members were tragically killed. The accident report by the Bureau of Enquiry and Analysis for Civil Aviation Safety concluded that the crash resulted because the co-pilot adjusted the auto-pilot settings so that the Airbus A320-211 descended until it impacted with the ground. *See*, Bureau of Enquiry and Analysis for Civil Aviation Safety, 'Final Report', March 2016, https://www.bea.aero/uploads/tx_elydbrapports/BEA2015-0125.en-LR.pdf.

10 Metrojet Flight 9268 was an international chartered passenger flight from Sharm El Sheikh, Egypt to Saint Petersburg, Russia. On 31 October 2015, the Airbus A321-231 disintegrated above Egypt's Sinai Peninsula tragically killing all 217 passengers and 7 crew on board. The Russian Federal Security Service, on 16 November 2015, reported that the crash was caused by a terrorist attack. The so-called 'Islamic State of Iraq and the Levant' claimed responsibility for the attack.

11 Aviation Safety Network, 'ASN Data Show 2017 Was Safest Year in Aviation History', 30 December 2017, https://www.news.aviation-safety.net/2017/12/30/preliminary-asn-data-show-2017-safest-year-aviation-history/.

12 *See*, Aviation Safety Network, 'Safety Indicator 2018', https://www.aviation-safety.net/statistics/period/barometer.php.

both which resulted in a tragic loss of lives, security aspects, while seen as separate in aviation, are linked to safety and play an important role in determining the success of the aviation industry.

Cybersecurity in aviation – a transversal topic – has made headlines, with a notable example taking place in 2015, when Chris Roberts, a cybersecurity consultant, was detained by the Federal Bureau of Investigation (FBI) in April 2015 following a United Airlines flight to New York. The arrest happened after officials saw Twitter posts in which Roberts claimed that he had hacked into the aircraft in which he was travelling on board. Roberts told the FBI that he had hacked into the in-flight entertainment systems on board around 15 to 20 aircraft from 2011 to 2014 and was able to control aircraft engines during those flights. Roberts' claims have not yet been verified.[13]

The Roberts' case is just one example of how technology by an unauthorised user can affect an aviation stakeholder. There are, however, many more examples, which this work will examine, and new previously unthought of cyber threats will become apparent as technology advances. The effects of the reported cybersecurity threats in aviation have been relatively minimal so far, however, the type of cyberattack and the motive behind it is varied and often unpredictable.

> Cybersecurity vulnerabilities have the potential to jeopardise civil aviation safety and efficiency.[14]
> Currently, the growing threat to keeping the aviation industry safe and secure from attacks lies in cyberspace.[15]

Cybersecurity in aviation is becoming a serious issue that all aviation stakeholders must consider in order to protect passengers, third parties and themselves.

In order to increase cybersecurity, regulators at all levels are beginning to react to the threat of cyberattacks. Regulatory reactions include those at an international level, such as an amendment to ICAO Annex 17, and those that specifically address cybersecurity at a regional level, which includes policy pushes from the European Union (EU) and the creation of specific rules.

This book addresses the question whether the current regulatory approach in the European Union is appropriate for international civil air transportation.

13 *See*, Chapter I.2.3.4 and Chapter I.3.2.
14 UK Centre for the Protection of National Infrastructure (CPNI), Security in Civil Aviation, August 2012.
15 *Ibid.*

Based on a critical analysis of EU aviation law, as well as related international law, with particular emphasis on the new regulatory developments in EU aviation safety, it will be argued that the current legal status quo is not appropriate and needs to be changed.

While the book will advocate and justify the need for a multilateral approach to aviation cybersecurity regulation, it will critically challenge the current way that multilateralism is being conducted. This is because it is happening in a fragmented way where information sharing is rare, secrecy is high, duplication is present and knowledge gaps are prevalent. Such a regulatory system, where cybersecurity is only as strong as the weakest link, is simply not suitable for a safe and secure aviation environment. Therefore, to challenge the current and continuous regulation of cybersecurity, the book will also assess six basic assertions that are found in the current literature that are relevant for both aviation and cybersecurity more generally. These are:

1. Cybersecurity is not one homogenised topic, but is an overarching title that encompasses numerous acts, different actors, varying victims and a plethora of consequences.
2. Regulation should not be the only tool in promoting cybersecurity.
3. The role of the European Union in aviation cybersecurity will increase.
4. Regulators must work more closely with stakeholders from all domains and not just aviation stakeholders or Member States.

The Convention on International Civil Aviation of 1944 and its 19 Annexes have not gone far enough to effectively and efficiently regulate aviation cybersecurity.

I KEY OBSERVATIONS IN AVIATION CYBERSECURITY

1 GROWTH OF AVIATION CYBERATTACKS

1.1 Introductory Remarks

To highlight the growing importance of cybersecurity in civil aviation, it is crucial to document the frequency of cyberattacks against aviation stakeholders.

> Cybersecurity: The measures taken to remove or protect vulnerabilities against cyberattacks, or mitigate the damage once a cyberattack has happened.[1]
>
> Cyberattack: A cyberattack is an attempt – whether successful or not – by an attacker to gain access to system services, resources, data or information, or an attempt to compromise system integrity. This includes both electronic and physical, and can compromise a single act or multiple via numerous methods.[2]

This is because, if there are no cyberattacks taking place or if they are taking place on a very limited scale with negligible consequences, such as to safety, security or commercial interests, then the relevance of this work is minimal. Whereas, if the opposite is found, then it highlights the need to assess whether the current legal environment is sufficient, thus providing preliminary justification for this topic to be tackled in a book. Therefore, this subchapter will list the publicly reported cyberattacks against aviation stakeholders and then try to understand why such attacks are taking place.

1.2 Genesis of Aviation Cyberattacks

The first recorded cyberattack against an aviation stakeholder took place on 10 March 1997 and was targeted against a telephone company that serviced, among others, Worcester, Massachusetts airport in the United States (US). The attacker identified the telephone number of different modems belonging to the telephone company, which granted the individual access and allowed a series of computer commands to be initiated. The

1 Also, 'cyber security' and 'cyber-security'.

2 Also, 'cyber attack' and 'cyber-attack'.

cyberattack resulted in the phone system at the airport crashing for approximately six hours. In total, the crash caused the Federal Aviation Administration (FAA) control tower, airport security, airport fire department, weather service provider and air carriers at the airport to be unable to use their phone services. Furthermore, the control tower's radio transmitter, as well the transmitter responsible for activating the runway lights, were also shut down. Finally, around 600 homes in the neighbouring town were also left without phone services. Although the cyberattack caused disruption, which resulted in delays, it was claimed that there was no risk to safety. The cyberattack prompted the Area Manager of the airport's air traffic control company to state that "[w]e dodged a bullet that day".[3] The attacker was later arrested, however, as a minor, his age, identity and address were not disclosed. This genesis case also marked the first time the US federal government brought a case against a juvenile for the commission of a computer crime. The case did not go before court and was instead settled via a plea agreement.[4]

> This case reflects our intention to prosecute in federal court anyone, including a teenager, who commits a serious computer crime. The plea agreement is a balanced effort, weighing the seriousness of this juvenile's computer intrusions and his lack of malevolence.[5]

It was reported that the plea involved two years' probation without the use of a computer, a fine and 250 hours of community service.

1.3 Catalogue of Aviation Cyberattacks

Luc Tytgat from EASA stated in 2016 that aviation systems are subjected to an average of 1,000 cyberattacks each month. Whether this number is correct is debateable as, for example, not every cyberattack is publicly reported. This may be for security or commercial reasons, or simply because the victim is unaware of the cyberattack, as it takes on average 100 days for an aviation stakeholder to discover a cyberattack. Furthermore, reporting figures also depend on the definition placed on 'cyberattack' and 'aviation system'. A generic phishing scam sent to thousands of email addresses, including an employee of an airline, may constitute, under some definitions, as a cyberattack on an aviation system, whereas other definitions may hold the cyberattack to be too general to be placed under the 'aviation' heading. Further, a bomb threat via Twitter may constitute a cyberattack as it involves the use of computer systems, as opposed to a threat delivered by physical mail,

3 CNN, 'Teen Hacker Faces Federal Charges', 18 March 1998, http://www.edition.cnn.com/TECH/computing/9803/18/juvenile.hacker/.

4 US Department of Justice, 'Juvenile Computer Hacker Cuts Off FAA Tower at Regional Airport – First Federal Charges Brought Against a Juvenile for Computer Hacking', Press Release, 18 March 1998.

5 *Ibid.*

or the involvement of technology may be too minimal to render it a cyberattack. In this work, a broad understanding of both 'aviation' and 'cyberattack' will be used in order to be as inclusive as possible at this exploratory stage. This may, however, not be a suitable approach for regulators. Considering all of this, all attempts were made to make the list below complete, by using public sourced information, but as the real extent of cyberattacks against aviation is unknown, there are substantial gaps.

Table 1 Publicly/Openly Reported Confirmed and Unconfirmed Cyberattacks

Date	Victim	Attacker	Attack Method	Result
Mar. 1997	Worcester Airport	Unidentified Minor	Targeted Hack	Loss of airport communication services.
Jun. 2001	Lufthansa	Andreas-Thomas Vogel	DDoS Attack	Lufthansa Internet homepage temporarily unavailable.
2006	Northwest Airlines	Christopher Soghoian	Fake website	A website provided fake boarding passes. This allowed the holder to access points in an airport beyond airport security.
Jul. 2006	FAA	Unknown	Malware – Virus	Several ATC systems in Alaska were shut down.
Aug. 2008	Barajas Airport	Unknown	Malware – Trojan Horse***	Total loss of JK5022.
Feb. 2009	FAA	Unknown	Malware**	Unauthorised access to 48,000 FAA personnel files.
2010-2016	South Korea	North Korea**	Jamming and Spoofing	GPS interference.
Mar. 2011	South Korea	North Korea**	DDoS Attack	Interference of passport, baggage and check-in systems.
Jun. 2011	Indira Gandhi Airport	Unknown	Malware**	Check-in services at airport disrupted.
Oct. 2011	US Military	Unknown	Malware – Keylogger	Potential compromise of sensitive information.
Dec. 2011	US Military	Iran	Spoofing	Total loss of an unmanned aircraft and loss of on board data.
2011-2014	United Airlines	Chris Roberts	Unauthorised Access**	Reported unauthorised control of aircraft while in flight.
2012	Newark Liberty Airport	Gary Bojczak (Unintentional)	Jamming	Interference with the ground-based augmentation system.
Jul. 2012	N/A – Experiment	Todd Humphreys	GPS Spoofing	Simulated forced landing of an unmanned aircraft.

Date	Victim	Attacker	Attack Method	Result
Aug. 2012	Unidentified non-US Airport	Unknown	Malware – Trojan	Potential compromise of sensitive information.
2013	75 US Airports	Unknown	Phishing	Sensitive information was compromised.
Apr. 2013	American Airlines	Unknown	Targeted Attack**	Loss of ground computer systems and delays.
Jul. 2013	Ataturk Airports	Unknown	Malware**	Passport control system crash and delays.
Jul. 2013	Sabiha Gokcen Airport	Unknown	Malware**	Passport control system crash and delays.
Oct. 2013	Maine Airport	Unknown	Jamming	Access blocked to Unicom frequency.
Mar. 2014	Malaysia Airlines	Unknown	Spoofing**	Potential total loss of MH370.
Apr. 2014	American Airlines	Unidentified Minor	Bomb Hoax via Twitter	None reported.
Dec. 2014	16+ States	Iranian Based Hackers**	Targeted Attack	Effects to airports and airlines in concerned States.
Feb. 2015	American and Delta	Unidentified Minor	Bomb Hoax via Twitter	Delays to flights.
Mar. 2015	British Airways	Unknown	Unauthorised Access	Frequent flyer information compromised.
Mar. 2015	Air France	Moujahidin Team	Unauthorised Access	Website defaced.
Apr. 2015	FAA	Unknown	Malware	None reported.
Apr. 2015	Lufthansa	Unknown	Botnets	Frequent flyer information compromised.
Apr. 2015	Ryanair	Chinese Based**	Hack	Loss of USD 5 million.
Jun. 2015	LOT	Unknown	Hack**	Loss of computer airlines ground systems.
Nov. 2015	Swedish CAA	Russian**	Targeted Attack	Radar disruption.
Jan. 2016	Boryspil Airport	Russian**	Malware	No effects reported.
Feb. 2016	Steven Petrow (American Airlines)	Unknown	Hack**	Passenger hacked other passengers' computers via on board WiFi.
Feb. 2016	Ryanair	Brian Lake	Twitter	Tweet claiming Ryanair would be subject to a terror attack.

Date	Victim	Attacker	Attack Method	Result
May 2016	British Airways	Paul Dixon	DDoS	Website crashed causing GBP 100,000 damage.
Jul. 2016	Vietnam Airports and Airlines	1937CN	Hack**	Information screens and sound systems showed anti-Government information.
Aug. 2016	Delta Airlines	Unknown	Hack**	Airline check-in services disrupted and delays.
Sep. 2016	Boeing 757	US Department of Homeland Security	Unauthorised Access via Radio Frequencies	Unreported.
Nov. 2016	Saudi CAA	Unknown	Malware – Virus	Computer damage that erased critical data and stopped operations.
Dec. 2016	Sialkot Airport	Mallu Cyber Soldiers	Unauthorised Access	Distribution of airport's log-in details.
Dec. 2016	Cochin and Thiruvananthapuram Airport	Kashmiri Cheetah of Pak Cyber Attackers	Server Hack	Manipulation of websites and potential compromise of sensitive information.
Jun. 2017	DLA Piper (Law firm with aviation services)	Russian Based Attackers**	Ransomware	No access to telephones and emails for several days.
Aug. 2018	Aircraft WiFi	Ruben Santamara	Unauthorised Access	Gain access to passengers' technologies.
Oct. 2018	Cathay Pacific	Unknown	Data Breach	Hackers accessed customers' personal information.

Note: All cited cases of cyberattacks against aviation stakeholders in this study can be found in Table 1.
* The website has since been taken down (https://www.dubfire.net/boarding_pass/).
** *Unconfirmed.*
*** *Later concluded not to be a cyberattack.*

Observing Table 1 reveals additional limitations:

1. Not every case can be confirmed or rejected as a cyberattack, as there is currently not enough available information.
2. Important information may become available later, which can determine or reject the case as a cyberattack.
3. Determination may also be made by third parties, such as experts, which make evidence-based conclusions, so the reliability of the conclusions is questionable.

Therefore, the above list will evolve over time, and caution should be given when assessing new and existing examples. Therefore, the minimal reports found in 2018, the final-complete year of this study, will rise. This is supported by the publicly restricted reports published by different aviation cybersecurity specialists.

Due to the difficulty in determining whether an event is a cyberattack or a technical fault, it is now common practice by the media to connect any computer-based issue with the possibility of a cyberattack. For example, on 27 May 2017, British Airways suffered a major information technology (IT) failure that affected over 1,000 flights causing delays and cancellations. A report from The Guardian newspaper explicitly stated that there was "no evidence of a cyber-attack".[6] Therefore, there is a practice to link or unlink a technical failure to a cyberattack, despite a lack of evidence either way, so caution also must be given when assessing initial reports.

1.4 Reasons for the Growth in Aviation Cyberattacks

Table 1 has shown that there is an increase in the number of cyberattacks taking place against aviation stakeholders, as compared with a decade ago. While micro factors, such as an attacker's motive, are essential for the conception of each individual cyberattack, these do not necessarily explain the overall trend. This study, thus, puts forward five macro causes for the overall increase in cyberattacks against aviation stakeholders.[7]

1.4.1 Increased Number of Computers

Antonio Chialastri, in his work *Automation in Aviation*,[8] identified three generations of aviation automation technology.

1. Mechanical.
2. Electrical.
3. Electronic.

At the birth of manned aviation, there were very few instruments on board an aircraft, with just the bare minimum to ensure flight. For example, navigation was performed by way of visual line of sight (VLOS) and paper maps, rather than with radar and a Global Navigation Satellite System (GNSS). As technology advanced, spurred by the realisation

6 H. Siddique, 'British Airways: Turmoil Continues after IT Failure Grounds Flights', The Guardian, 28 May 2017, https://www.theguardian.com/business/2017/may/28/british-airways-cyber-attack-unlikely-amid-scramble-to-resume-flights-on-sunday.

7 A part of Chapter I.1.4 is based on the research carried during the writing of S. Hobe and B. Scott, 'International Civil Aviation and the Dehumanisation of Activities', 60 *German Yearbook of International Law* 144-173 (2017).

8 A. Chialastri, 'Automation in Aviation', *in* Florian Kongoli (ed.), *Automation*, First Edition, (InTech, 2012).

of the utility of aviation, more mechanical instruments were introduced. As aircraft became larger and more complex, the need for mechanical support became increasingly necessary. This initial *mechanical* stage began the first steps of widening the "gap between the pilot's input (action on the yoke) and the final outcome (aerodynamic movement)".[9]

The general advancement in electric technology allowed this to enter slowly into aviation, which saw the replacement of mechanical with electrical devices. The second generation of *electrical* took-off in the 1960s, which included:

> electric autopilots, auto-throttle (to manage the power setting in order to maintain a selected speed, or a vertical speed), flight directors (used to show pilots how to manoeuvre to achieve a pre-selected target such as speed, path-tracking and so forth), airborne weather radars, navigation instruments, inertial platforms, but also improved alarming and warning systems capable of detecting several parameters of engines and other equipment.[10]

The third generation was predominantly driven by the *electronic* revolution occurring from the mid-1980s onwards.

> Electronics significantly helped to diminish the clutter of instruments on board and allowed for replacing old indicators – gauges in the form of round-dial, black and white mechanical indicators for every monitored parameter – with integrated coloured displays [...] capable of providing a synthetic and analytic view of multiple parameters in a limited area of the cockpit.[11]

Therefore, the third generation can be marked as the introduction of computers. Thus, this generation installed a host for cyber vulnerabilities, such as computers, into the aviation sector that can be exploited by attackers.

Vulnerability: A flaw or weakness in a technology (information system), either physical or electronic – as well as a system security procedure, internal control or implementation – that is susceptible to exploitation via a cyberattack.[12]

Although the introduction of electronics into aviation was initially small, it is now impossible to carry out international air passenger transport without the assistance of com-

9 *Ibid.*, at 85.
10 *Ibid.*
11 *Ibid.*, at 86.
12 *See*, for a vast list of unpatched vulnerabilities, Common Vulnerabilities and Exposures, 'Home', https://cve.mitre.org/. This does not take into account zero days.

puters and related systems. These include those installed on an aircraft, ground infrastructure at airports, online ticket sales and the running of offices. The advancement of electronics is also seen with air traffic control (ATC) and air navigation services (ANS), ground handling and other stakeholders.

1.4.2 Increased Connectivity with the Internet

The problem has been exacerbated by the introduction and advancement of the Internet of Things (IoT). This refers simply to the connection of devices to the Internet. This has come to include devices, such as cars, street lighting, fridges, watches and aircraft. It has been predicted that IoT devices will total more than 24 billion by 2020.[13] This would equate to USD 6 billion of investment in IoT solutions, including application development, device hardware, system integration, data storage, connectivity and cybersecurity, which will produce around USD 13 trillion by 2025.[14] The aviation sector is already investing in the IoT.

> Many airlines have already caught on to this potential. A recent survey from SITA found that an impressive 86% of airlines think the Internet of Things will lead to clear and significant benefits for them and their customers. That promise is already being put into action for some as the same survey showed that 37% of airlines are already budgeting for the technology. Beyond the current strategies used, 16% of airlines plan to start major IoT programs by the year 2018, with 41% expecting to do more extensive research and development into IoT in the next few years. It's clear that in whatever form it takes, the Internet of Things will have a major role to play for airlines and airports.[15]

Some aviation sector examples of IoT include:

1. Virgin Atlantic's Connected 787: Virgin Atlantic is investing in Boeing 787 aircraft that utilise the IoT to generate half a terabyte of data per flight. The purpose behind this is to assess the health and performance of the aircraft so, for example, it can identify and fix mechanical problems straight away, thus increasing safety and reducing delay.
2. Helsinki Airport: Helsinki Airport in Finland is using WiFi and iBeacons[16] to track passengers and offer location-based services. By doing this, the airport can prevent

13 A. Meola, 'What Is the Internet of Things (IoT)?', Business Insider, 19 December 2016, http://www.businessinsider.com/what-is-the-internet-of-things-definition-2016-8.

14 *Ibid.*

15 R. Delgado, 'How Airlines Are Taking Advantage of the Potential of the Internet of Things', IoT News, 18 July 2016, https://www.iottechnews.com/news/2016/jul/18/how-airlines Aviation Stakeholder:-are-taking-advantage-potential-internet-things/. *See*, SITA, 'Airline IT Trends Survey 2015', 18 June 2015.

16 *See*, Apple, 'iBeacon', https://developer.apple.com/ibeacon/.

queues and bottlenecks, and retailers can send push notifications to travellers based on their exact location.

3. Delta's Baggage Tracking: Delta uses Radio Frequency Identification baggage tracking technology. Via push notifications on Delta's mobile application, passengers can see the location of their baggage throughout the journey.
4. Rolls Royce: The engine manufacturer partnered with Microsoft to diagnose potential faults in engines fitted to aircraft. Information on engine health, air traffic control, route restrictions and fuel use will be collected from hundreds of sensors inside the engines, and analysed to detect any operational anomalies or signs of developing faults.[17]

Manufactures, airlines and airports are utilising IoT to improve safety, reduce costs and increase efficiency.

> The cyber world of interconnected and interdependent systems has increased the vulnerability of aircraft and systems and therefore the potential impact that breaches in security can have. More attention is therefore due to this complex but containable problem.[18]

There are many more examples in aviation and it is likely that as technology advances and as more entities begin to see the benefits of IoT, new and more creative ways of using it will become apparent. However, the more technologies that are connected to the Internet, the more items that are becoming vulnerable to a cyberattack.

Not only are devices becoming connected to the internet, they are also being connected to other devices. This, therefore, adds an additional layer to the connectivity of devices, which produces more vulnerabilities.

1.4.3 Increased Automation

Automation[19] in aviation is prevalent and a component in all of the main activities of aviation stakeholders. Some of the more well-known tools used in aviation, which provide some context, are:

17 *See*, D. Drinkwater, '10 Stellar Real-Life Examples of IoT Taking Flight in Aviation', Internet of Business, 3 August 2016, https://internetofbusiness.com/10-real-life-examples-iot-aviation/; and M. Drummond, '5 Great Ways Airlines Are Using the Internet of Things', Mercator, 18 August 2016, https://www.mercator.com/blog/5-great-ways-airlines Aviation Stakeholder:Airlines-are-using-the-internet-of-things.

18 UK Centre for the Protection of National Infrastructure (CPNI), Security in Civil Aviation, August 2012.

19 Automation may become supplemented by artificial intelligence (AI).

1. Auto-Pilot: This is a system used to control the trajectory of an aircraft without constant 'hands-on' control from the pilot. Autopilot does not replace a human operator, rather it assists in controlling the aircraft, allowing the pilot to focus on other elements of the flight.
2. Auto-Landing: This is a fully automated system that performs the landing procedure, whereby the flight crew takes a supervisory role. This system allows an aircraft, for example, to land in bad weather conditions, such as fog and heavy rain, which would otherwise be difficult or impossible for a pilot to operate.
3. Traffic Collision Avoidance System (TCAS): This on board system detects other aircraft in the area that could pose a collision hazard and then selects a mutually agreeable avoidance manoeuvre for the concerned aircraft. The flight crews must follow the instructions so as to avoid the collision.
4. Enhanced Ground Proximity Warning System: This is a similar system to TCAS, which is designed to prevent collisions with the terrain.[20]
5. Airport Baggage Systems: Airports often have fully integrated automated high-speed airport baggage handling systems. These systems include security screening, storage, allocation and transportation of baggage (e.g. from check-in to the departure gate, transfers, and aircraft to arrival hall).[21]
6. Unmanned Aircraft: These types of aircraft can rely on automatic technologies to operate, such as the use of follow-me mode Detect and Avoid (DAA).

It can be seen from the above examples that automation technology is not always for the use of one stakeholder, such as TCAS, which has a useful application for the airlines, the people on board of the aircraft, other airspace users and air traffic controllers. Furthermore, automation technology can be used for a variety of outcomes, such as for safety, efficiency gains, environmental protection, commercial benefits and security. However, what automation does provide across the board is reliance on that technology. This supports the growth in computers and can provide a suitable environment for attackers as automation can increase the number of vulnerabilities. This could be through the increased use of computers and the Internet, the limited authorised human supervision, potential compliancy due to the hands-off approach by users or ignorance as to how the technology should function.

20 A. Pritchett, 'Aviation Automation: General Perspectives and Specific Guidance for the Design of Modes and Alerts', 5(1) *Reviews of Human Factors and Ergonomics* 82-113 (2009).
21 S. Hobe and B. Scott, 2017.

1.4.4 Increased Motives

A cyberattack stems from a particular motive. The motive may be to make a political statement, extract information or money, seek revenge, spy on the competitor, or hinder a State. Due to the importance of aviation for global trade, the international movement of people, gross domestic product (GDP), among other things, this sector lends itself to many motives. For example, aviation:

1. stakeholders are sometimes State owned, such as airports and airlines, so targeting them could be for political reasons;
2. is often seen as safe, so a successful cyberattack could produce panic;
3. is an important sector for countries and industry, so a cyberattack can have severe consequences;
4. can be a very public target as it may be hard to hide the consequences of a cyberattack;
5. stakeholders deal with many passengers, some of which are stressed or delayed, so revenge cyberattacks cannot be ruled out; and
6. stakeholders often have a lot of capital and hold a lot of personal data, such as passenger credit card details, and there is thus a strong financial incentive to attack these entities.

As a result, the aviation sector presents numerous and varying motives to attackers, which contributes to the growing number of cyberattacks.

1.4.5 Increased Reliance on Cyberattacks

Cyberattacks are becoming the preferred tool for some individuals and groups. This may, *inter alia*, be a result of:

1. the growing connectivity in the world, so once non-existent vulnerabilities are now accessible;
2. the ease that some attacks can be implemented as demonstrated by the availability of off-the-shelf malware;
3. the fact that attacks can be initiated remotely so anonymity can be achieved; or
4. the ability to avoid prosecution as many legal systems are not equipped to deal with these threats.

Whatever the reason, cyberattacks are becoming a preferred method of attack over more traditional methods. This is not only a point for aviation, but for other industries.

1.4.6 Putting It Together

What can be concluded is that aviation is increasingly relying on computers in all aspects. These computers are increasingly being connected to the Internet and other networks, as well as becoming more automated. Therefore, the aviation sector is producing more vulnerabilities that can be exploited by an attacker. In addition, the aviation sector is an appealing target. As a result, there are many reasons why aviation is being subjected to a growing number of cyberattacks. None of the reasons highlighted in this subchapter are likely to cease, so this problem is unlikely to fade away.

The findings and conclusion in this subchapter are not unique to aviation. For example, other transportation sectors, like space, road, rail and water, are becoming more reliant on computers, the IoT and automation. This is also the case for other sectors like finance and energy. Therefore, while the growth in numbers of cyberattacks can be explained by an analysis of aviation, the causes are not unique to this sector and quite simply, aviation is just another target. However, the consequences may be quite different.

2 Aviation Cyber-Based Attack Methods

2.1 Introductory Remarks

As Table 1 shows, cyberattacks do not involve only one form of attack method. Rather, numerous attack methods are available to an attacker, which have different stages of implementation and target different vulnerabilities. This subchapter will analyse the core electronic and physical attack methods that have been initiated against aviation stakeholders in the past.

> Electronic Attack Method: A cyberattack that is implemented using a digital technology, such as a laptop or smartphone.
>
> Physical Attack Method: A cyberattack that is implemented without using a technology, but has a role in a cyberattack or has consequences for the victim's cybersecurity.

For a cyberattack to be carried out, it requires the use of an attack method. The core attack methods, as used against aviation stakeholders, will be analysed below.

2.2 Electronic-Based Cyberattack Methods

2.2.1 Malware

Malware, which is a portmanteau for 'malicious software', is an umbrella term that refers to a variety of hostile or intrusive software. Under this umbrella, several different attack methods can be identified.

Adware: Software that introduces advertising banners to the programme that is running.

Botnet: A number of electronic devices, mostly computers ('bots'), which includes malicious software. Botnets not only infect other devices but can also implement other attack methods like DDoS attacks.

Cryptojacking: This is an emerging cyber threat, which uses a technology's resources to 'mine' cryptocurrencies.

Keylogger: A programme that records the keyboard keys that are pressed.

Ransomware: Software that restricts access to the victim's computer or files, and then displays a message demanding payment in order for the removal of the restrictions to occur.

Spyware: Software that collects the victim's data or information stored on the device which is infected.

Trojan Horse: A programme that appears to perform a desired function, but carries a hidden programme.

Virus: A software programme that replicates itself, once it is executed, by modifying other computer programmes and introducing its own code.

Worm: A computer programme that replicates itself to spread to other computers.

The impact of malware is varied as it can range from the slowing down of equipment, to the loss of sensitive data. The risk is intensified as, while it will be known when a virus prevents the victim from using the device (active), it may not be observable when data-collecting malware is present as it may remain hidden (passive). The consequences can impact either tangible property (e.g. denied access to a computer) or intangible property (e.g. intellectual property rights).

Active cyberattack: An attack that is disruptive, such as slowing down use, deleting data or restricting access, usually making the victim aware.

Passive cyberattack: An attack that is non-disruptive, such as information gathering, resulting in the victim sometimes being unaware.

A major example of this was the Key Logging Virus that affected the US military unmanned aircraft fleet at Creech Air Force Base in Nevada in September 2011, which recorded the remote pilot's actions.[22] The effects of each depend on their designed intention; they could simply lay dormant with no noticeable effect or could cause the total loss of the system.

2.2.2 Denial of Service and Disruptive Denial of Service

A Denial of Service (DoS) attack is where

> one computer and one Internet connection is used to flood a server [...]. The goal of such a denial of service attack is to overload the targeted server's bandwidth and other resources. This will make the server inaccessible to others, thereby blocking the website or whatever else is hosted there.[23]

A Disruptive Denial of Service (DDoS) uses more than one computer and Internet connection, sometimes being situated in different locations. A DDoS has a higher chance of overloading the victim.[24]

The cyberattack on Lufthansa in June 2001 is an example of such a DDoS attack.[25] A common method is to send fake requests to the victim's website to exhaust the server resources and this then stops the legitimate use of the website. This could have a profound effect on aviation as it could prevent passengers from checking-in online prior to flying, monitoring flight information or purchasing tickets.

2.2.3 Jamming and Spoofing

Jamming refers to the transmission of electronic signals that overwhelm the signals of a specific sender or receiver. This denies service to authorised users because the legitimate traffic is jammed by the harmful frequencies of the illegitimate traffic.[26] This has been witnessed in South Korea within the context of airports and unmanned aircraft. However, this is by no means limited to ground infrastructure or unmanned aircraft as, for example, the communication or navigation facilities of a civil manned aircraft could be blocked

22 N. Shachtman, 'Computer Virus Hits US Drone Fleet', Wired Magazine, 7 October 2011, http://www.wired.com/dangerroom/2011/10/virus-hits-drone-fleet/.

23 L. Munson, 'DoS Vs DDoS – What is the Difference?', Security-FAQs, 29 May 2011, http://www.security-faqs.com/dos-vs-ddos-what-is-the-difference.html.

24 *See*, for more specific examples, TCP SYN flood attacks, teardrop attacks, smurf attacks and ping of death attack.

25 *See*, Chapter I.3.3.

26 *See*, for more specific information, Man in the Middle attacks, session hijacking, IP Spoofing and Replay.

or interfered with, so that the operator becomes blind and the aircraft will either crash, initiate its emergency procedures or land.[27] This was the case at Newark Liberty Airport in 2012 where Gary Bojczak unintentionally jammed ground-based systems at the airport with a USD 100 Global Positioning System (GPS) jammer and this prevented aircraft from taking-off and landing.

Spoofing is similar to jamming, but it is more invasive as it actively interferes. This is because, in addition to blocking the signal, it also manipulates the victim's technology.

Technology: This is any device, including hardware, and software.

Spoofing in aviation could include creating false civil GNSS signals that trick the aircraft's GNSS receiver into thinking that nothing is wrong. The false signal, as introduced by the attacker, could then guide the aircraft along a new navigational course.[28] A US unmanned aircraft which was seized by Iranian forces is reported to have been brought down by spoofing. Another notable example of spoofing is within the context of satellites whereby, in 2007, a Sri Lankan Tamil rebel organisation hijacked a transponder on a commercial satellite and used it to broadcast pro-rebel propaganda.[29]

2.2.4 Tapping and Eavesdropping

Tapping is the unauthorised access to recorded information. This can involve cracking encrypted communication links which may require sophisticated and time-consuming methods, or it may simply require tuning into the right frequency or bandwidth. For example, many of the US Reaper and Predator military unmanned aircraft broadcasted classified video streams to personnel on the ground 'in the clear'. In other words, the information was broadcasted without encryption. In 2009, the US discovered "days and days and hours and hours" of unmanned aircraft footage stored on the laptops of Iraqi insurgents.[30] The tapping was made possible with the aid of a USD 26 piece of software. Eavesdropping is the unauthorised real-time interception of private communications, such as a phone call, instant message, videoconference or fax transmission.

27 Infosec Institute, 'Hacking Drones … Overview of the Main Threats', 4 June 2013, https://www.resources.infosecinstitute.com/hacking-drones-overview-of-the-main-threats/.

28 Aerospace Engineering and Engineering Mechanics, 'Todd Humphreys' Research Team Demonstrates First Successful GPS Spoofing of UAV', 28 June 2012, https://resources.infosecinstitute.com/hacking-drones-overview-of-the-main-threats/Interference:Spoofing.

29 P. Selding, 'Intelsat Vows to Stop Piracy by Sri Lanka Separatist Group', Space News, 18 April 2007, http://www.spacenews.com/intelsat-vows-stop-piracy-sri-lanka-separatist-group/.

30 N. Shachtman, 2011.

2.2.5 Phishing

Phishing is the attempt to obtain sensitive information, such as credit card details, usernames and passwords, often for malicious reasons, by the attacker who is disguised as a trustworthy source. Phishing is typically carried out by email or instant messaging, and it often directs users to enter personal information into a fake website, which is identical or similar to the legitimate one. Communications purporting to be from social media websites, banks, online payment processors or IT administrators are often used to lure victims. Phishing is an example of social engineering techniques used to deceive users, and exploits weaknesses in current web security.

Ryanair customers experienced a phishing cyberattack in Summer 2017. A Facebook post and online survey-scam stated that Ryanair was offering free flights due to the mass cancellations resulting from pilot scheduling errors. The offer was fake and was an attempt by attackers to gain personal information.

Phishing, as well as other types of attack methods, may use social engineering to implement the cyberattack.

> Social engineering: This is the psychological manipulation of people into performing specific actions or getting them to divulge certain information.

Social engineering can utilise either electronic, such as emails, or physical, such as face-to-face conversations, in order to achieve the intended outcome.

Phishing also has subcategories. First, spear-phishing is targeted phishing geared towards specific individuals or companies. Second, phishing cyberattacks that specifically target senior executives within an organisation have been termed 'whaling attacks'.

2.3 Physical-Based Cyberattack Methods

2.3.1 Theft

Physical technology may be stolen and any data stored on that technology will also subsequently be stolen. The unauthorised acquisition of this data could lead to cybersecurity risks as it could be used and disseminated. In addition to the loss and unwarranted use of the technology and information, the equipment can also be studied. This would allow the technology's vulnerabilities to become known so that it could be easier to subject it to a cyberattack in the future.

2.3.2 Tampering

Physical tampering is unauthorised interference with the technology. This could include fitting an aircraft with a tracking device, attaching hardware that can transmit any recorded information or installing a device that can allow an outside source to take over the system. This could also include tampering with the data stored on the technology. This would then affect the integrity and accuracy of the information.

2.3.3 Forceful Control

Forceful control is where an unauthorised person takes over by physical force. Examples of this could include people trespassing onto facilities and then taking control of a technology. This can have cybersecurity implications, for example, a cyberattack could then be launched from that technology, which is inside the victim's computer system, so passwords and firewalls are no longer an issue.

2.3.4 In Situ

It may be the case that a cyberattack takes place on location. An example of this is Roberts' alleged attack into in-flight entertainment systems and altering the altitude of the aircraft. This can also happen when the technology is being operated by authorised people, but it is being employed for unauthorised purposes. For example, an aircraft may be fitted with sensory equipment that is then used to illegally record sensitive information. An example of this happened in the United Kingdom (UK). On 25 August 2013, an unmanned aircraft was recovered from the waters located near to a nuclear submarine testing facility operated by the public limited defence company BAE Systems in Barrow-in-Furness, UK. Attached to the vehicle was a camera[31] and analysis of the video footage by the police revealed that the aircraft flew into the restricted airspace around the nuclear submarine facility.[32] The police then traced the unmanned aircraft back to Robert Knowles who was convicted for flying the unmanned aircraft in restricted airspace over a nuclear installation pursuant to Regulation 3(2) of the Air Navigation (Restriction of Flying) (Nuclear Installations) Regulations 2007.[33] While Knowles denies any intention, it does show that an unmanned aircraft can be *legally* fitted with technology and then used by a pilot to record sensitive data which is in violation of the law.

31 The retrieved footage is accessible on YouTube. *See*, YouTube, 'Walney Channel, Robert Knowles Prosecution Flight', 3 April 2014, http://www.youtube.com/watch?v=hriyv8WRHg4.

32 *See*, B. Scott, 'The First UK Conviction for the Illegal Use of an Unmanned Aircraft and How It Can Help Improve Regulations within the EU', 14(1) *The Aviation and Space Journal* 11-22 (2015).

33 "Subject to regulations 4 to 13, no aircraft is to fly over a nuclear installation to which this regulation applies below the height above mean sea level specified in Column 4 of the Second Schedule opposite its name." Air Navigation (Restriction of Flying) (Nuclear Installations) Regulations 2007, No. 1929 (2007), Section 3(2).

3 Aviation Cybersecurity Victims

3.1 Introductory Remarks

Highlighting the different victims of cyberattacks within the aviation industry is an important exercise. This is because it will determine whether such cyberattacks are being directed towards a single aviation stakeholder, thus demonstrating the simplicity of the topic in hand, or whether the situation is more complex involving numerous different stakeholders with different responsibilities and resources. Therefore, this subchapter will be dedicated to highlighting and discussing the different aviation stakeholder victims.

> Aviation Stakeholder: Any natural or legal person, group or other entity that has a link to air transport. This includes; airlines, airports, air traffic control providers, manufacturers, passengers, cargo delivery companies, ground handling, general aviation, unmanned aviation, States, national aviation authorities (NAA), regional and international organisations, and connected services (e.g. commercial entities, such as travel agents, catering services and contractors).
>
> Victim: Any aviation stakeholder that has had a cyberattack, whether directly or indirectly, or successfully or not, targeted against them.

The terms 'stakeholder' and 'victim' are defined, for the purpose of this work, in broad terms. This is to make the scope as inclusive as possible, in order not to unnecessarily exclude anyone or thing from the discussion. As the UK CAA states, "[a]viation affects so many people in so many ways, meaning we have to take the interests of a wide range of potential stakeholders into account".[34]

3.2 Airlines and Manufactures

The importance of computers for aviation is well documented for commercial aspects, navigation, and communications. However, computer systems are also necessary for the flight of most modern aircraft. The importance of this was made evident on 9 May 2015 when the engine control software on board an Airbus 400M, a four-engine turboprop military transport aircraft, was incorrectly installed, which resulted in it crashing and killing four of the six crew members. This was not a deliberate act and was instead attrib-

34 UK CAA, 'Our Stakeholders', https://www.caa.co.uk/Our-work/Corporate-reports/Strategic-Plan/Our-stakeholders/.

uted to human error, so it does not amount to a cyberattack, but it does show the devastating consequences that a computer failure can have.

Spanair flight JK5022, a Spanish domestic scheduled flight from Madrid's Barajas Airport to Las Palmas, crashed on 20 August 2008 immediately after take-off tragically killing 154 passengers and the entire six team crew. Only 18 passengers survived the crash and the subsequent fire on board the McDonnell Douglas MD-82. *El Pais*, a Spanish newspaper, reported that the crash was attributable to a Trojan Horse that infected computer systems, which prevented the computer system from functioning correctly.[35] This would then mark the first reported fatal cyberattack against civil aviation. Many media sources picked up on *El Pais*' report and this lead to wide distribution of the story.[36] However, it turned out to be media sensationalism as the official accident investigation report dismissed this claim, attributing the cause to pilot error, with the malware instead being at the airline's maintenance headquarters in Palma de Mallorca.[37] Thus, the malware was too remote to influence the safety of the flight.

There are also reports that Malaysia Airlines Flight MH370, a Boeing 777-200ER flight from Kuala Lumpur, Malaysia, to Beijing, China, which disappeared on 8 March 2014 with 239 people on board, may have been victim to a cyberattack against the on board systems which resulted in its disappearance.[38] However, such reports are currently unsubstantiated.

Another reported example of a cyberattack being directed towards the aircraft involved Roberts, a cybersecurity consultant, who was detained by the FBI in April 2015 following a United Airlines flight to New York, US. The arrest happened after officials saw Twitter posts in which Roberts claimed that he had hacked into the aircraft in which he was travelling. Roberts told the FBI that he had hacked into in-flight entertainment systems on board 15 to 20 aircraft from 2011 to 2014, and was able to control the aircraft engines during a flight.[39] The in-flight entertainment systems on board such commercial aircraft should be isolated from any essential flight facilities, so it is likely that Roberts' claim was more of an ill-thought joke than an actual infiltration.

---

35 J. Hernández, 'El Ordenador de SPANAIR que Anotaba los Fallos en los Aviones Tenia Virus', El Pais, 20 August 2010, http://www.elpais.com/diario/2010/08/20/espana/1282255211_850215.html.

36 *See*, as way of example, J. Leyden, 'Trojan-Ridden Warning System Implicated in Spanair Crash: Cascading Fail', The Register, 20 August 2010, https://www.theregister.co.uk/2010/08/20/spanair_malware/.

37 E. Bott, 'Fact Check: Malware Did Not Bring Down a Passenger Jet', Zdnet, 24 August 2010, http://www.zdnet.com/article/fact-check-malware-did-not-bring-down-a-passenger-jet/.

38 *See*, as way of example, SCADA/ICS Security, 'Cyber Threats Against the Aviation Industry', infosec Institute, 8 April 2014, https://resources.infosecinstitute.com/cyber-threats-aviation-industry/#gref.

39 *See*, as way of example, E. Perez, 'FBI: Hacker Claimed to Have Taken Over Flight's Engine Controls', CNN, 19 May 2015, https://www.edition.cnn.com/2015/05/17/us/fbi-hacker-flight-computer-systems/index.html.

Isolation may be achieved by retrofitting existing (legacy) aircraft or building it directly into new aircraft, which are designed to have the on board aircraft networks used by the airline completely isolated from passenger Internet services like WiFi.[40] Boeing has gone so far as to publicly deny the possibility of such a cyberattack.[41]

A US Department of Homeland Security official stated during his keynote address at the CyberSat Summit on 8 November 2017 that his team had successfully remotely attacked an aircraft. The attack took place on 21 September 2016 against a Boeing 757 that was parked at the Atlantic City airport in the US. There are four caveats to this claim:

1. The details of the attack have remained classified with the only statement being that the "team of industry experts and academics pulled it off by accessing the 757's 'radio frequency communications'."[42] Therefore, the consequences of the attack are unknown, such as whether the team was able to manipulate flight critical equipment. Information is also not known about how much resources were invested, both time and money, into the attack. Conclusions about the replicability in the real-world is consequently unknown.
2. The attack took place while the aircraft was stationary on the ground. It is unclear whether the attack could be replicated while the aircraft is in flight.
3. Boeing started production of the 757 in 1981 so this is an old aircraft. As a result, the same results may not transpire when newer aircraft are tested.
4. Boeing stopped producing 757s in 2004 so the amount of these aircraft is in decline. In July 2017, there were 689 aircraft in service out of the total 1,050 built by Boeing.[43] Consequently, the scope of this vulnerability will become less as more 757s are taken out of service.

Manufacturers' aircraft are not the only target as, for example, the companies themselves are also targeted. Stephane Lenco, the Airbus Group's Chief Information Security Officer stated to the Australian Cyber Security Centre conference that "[w]e get thousands of attacks daily because we have front ends to the Internet".[44] However, there are also serious cases, as Lenco further stated that "[w]e get on average 10 to 12 really, really serious

40 *See*, as way of example, FAA, 'Special Conditions: Boeing Model 787-8 Airplane; Systems and Data Networks Security-Isolation or Protection from Unauthorized Passenger Domain Systems Access', [E7-25467].

41 M. Trögeler, 'Collaboration Across the Global Aviation Community', European Aviation Club, 'Securing the Skies: A Framework for Aviation Cyber Security', Brussels, Belgium, 22 May 2017.

42 M. Smith, 'Homeland Security Team Remotely Hacked a Boeing 757', 12 November 2017, CSO, https://www.csoonline.com/article/3236721/security/homeland-security-team-remotely-hacked-a-boeing-757.html.

43 D. Thisdell and Rob Morris, 'World Airliner Census 2017', Flight International, 12 August 2017.

44 A. Coyne, 'How Airbus Defends Against 12 Big Cyber Attacks Each Year', IT News, 14 April 2016, https://www.itnews.com.au/news/how-airbus-defends-against-12-big-cyber-attacks-each-year-418131.

[attacks], that we would want to look at very deeply, a year".[45] Airbus, has been affected, for example, by ransomware which has denied access to infected computers.

This demonstrates that cybersecurity is not just a concern for the manufacturers' end products, but also for the day-to-day running of the company, whereby the whole supply chain is vulnerable.[46]

3.3 *Airlines and Ground Infrastructure*

While aircraft vulnerabilities have a direct link to both manufacturers and airlines,[47] airlines have also been subjected to cyberattacks against their ground infrastructure. American Airlines was forced to ground all flights for several hours on 16 April 2013 following a problem with its computer systems in the US. More than 400 flights were cancelled and many more were delayed following a four-hour system issue. American Airlines said the problem was caused by an inability to gain access to its reservations system called Sabre Global Distribution System, which is responsible for bookings and reservations, and also manages a wide variety of functions related to flights, including; printing boarding passes, online check-ins, ticketing and tracking bags. Sabre Holdings, which owns Sabre Global Distribution System, rebutted this stating that the problem did not stem from its computer systems. It has not been confirmed whether this was a cyberattack, however, similar incidents are becoming more common in the US. United Airlines faced a similar event on 14 October 2016 which resulted in around 60 flights being delayed. The effects of this incident were not as profound as with Delta as it took place during the night in the US. It has also not been confirmed as a cyberattack. However, the similarities between the Delta and United cases has led some cybersecurity experts to connect dots and conclude that they were both cyberattacks.[48]

One of the most prominent airline examples took place on 21 June 2016, where approximately 1,400 passengers of the Polish airline LOT were grounded at Warsaw Chopin Airport, Poland following a cyberattack against its ground computer systems used to

45 *Ibid.*

46 *See*, for more information about supply chain, as well as an early example of a cybersecurity report, AIAA, The Connectivity Challenge: Protecting Critical Assets in a Networked World. A Framework for Aviation Cybersecurity. An AIAA Decision Paper, August 2013, at 11. *See also*, E. Iasiello, 'A Review of 'A Framework for Aviation Cybersecurity' – A Decision Paper from the American Institute of Aeronautics and Astronautics', 11(1) *Homeland Security and Emergency Management* 95-100 (2014).

47 *See*, R. Goldberg, 'Cyber Risks Confronting Airlines: A Practical Approach to Manage the Risks and Pursue the Wrongdoers', 28(6) *Air and Space Law* 294-310 (2003).

48 J. Maxfield, 'Are America's Airlines Under Cyber Attack?', The Motley Fool, 30 January 2017, https://www.fool.com/investing/2017/01/30/are-americas-airlines Aviation Stakeholder:Airlines-under-cyberattack.aspx.

issue flight plans. The airport itself was reported not to be affected and LOT stated that there was no danger to passengers as the cyberattack did not affect any of LOT's system utilised by aircraft in flight.[49] Although the system was running again after five hours, this nevertheless resulted in domestic and international LOT flights being cancelled, with many also being delayed. Flights were back to normal on the following day. However, LOT rebooked passengers that were booked on cancelled flights and provided assistance to them, such as hotels, which resulted in financial and reputational loss. The steps taken by LOT result from its obligations contained in Regulation 261/2004.[50] Poland's domestic intelligence agency said it had been called in to investigate the incident and a spokesman for the airline stated that the cause of the problem was most likely attributable to a DDoS attack: "This was a capacity attack, which overloaded our network".[51] While DDoS attacks are fairly common and can require very minimal knowledge to implement, as most attacks are against publicly accessible Internet sites, such as the victim's homepage, here, the LOT system that was affected had no public site. This led one cybersecurity expert to comment: "I am quite surprised that such sensitive systems dedicated to airline operations are exposed to the Internet to be exposed to denial of service attacks".[52] The attacker has still not been located.

On 20 June 2001, human rights activists carried out a DDoS attack against Lufthansa's website which knocked out its Internet homepage for several minutes. The cyberattack was initiated by Libertad, a civil liberties group, which claimed that it was a protest against the profits made by Lufthansa for assisting the German Government with the deportation of up to 10,000 rejected asylum seekers each year. The demonstrators also disrupted Lufthansa's annual shareholders meeting in Cologne, attended by circa 5,000 shareholders, who had to be removed by the police. Lufthansa stated that it cannot be held responsible for the German Government's asylum policy and that it does not deport arbitrarily. One such case comes from 1999, when a Sudanese passenger died on a Lufthansa flight after allegedly suffocating in a special helmet that he was forced to wear. Consequently, the airline no longer carries deportees requiring restraints. Andreas-Thomas Vogel, who ran Libertad's website at the time of the event, was arrested following the event. A lower court in Frankfurt found Vogel guilty of using force against the airline.[53]

---

49 W. Szary and E. Auchard, 'Polish Airline, Hit by Cyber Attack, Says all Carriers are at Risk', Reuters, 22 June 2015, https://www.reuters.com/article/us-poland-lot-cybercrime/polish-airline-hit-by-cyber-attack-says-all-carriers-are-at-risk-iduskbn0p21dc20150622.

50 Regulation (EC) No. 261/2004 of the European Parliament and of the Council of 11 February 2004 Establishing Common Rules on Compensation and Assistance to Passengers in the Event of Denied Boarding and of Cancellation or Long Delay of Flights, and Repealing Regulation (EEC) No. 295/91, OJ L 46, 17 February 2004, pp. 1-8. *See*, B. Scott and A. Trimarchi, 'The Digital Aviation Industry: A Balancing Act Between Cybersecurity and European Consumer Protection', 4-5 *Air and Space Law* 443-462 (2017).

51 W. Szary and E. Auchard, 2015.

52 *Ibid.*

53 *OLG Frankfurt am Main*, 1 SS 319/05, 22 May 2006.

This was "based predominantly on the economic losses the airline had suffered during the campaign, both in terms of lost sales and the costs of acquiring additional bandwidth to soak the protesters' traffic".[54] The case was then appealed,[55] where the ruling was overturned because "the online demonstration did not constitute a show of force but was intended to influence public opinion".[56]

> The Vogel case was the first international precedent to recognize the legal and philosophical arguments put forth by supporters of DDoS activist actions. The court decision pivots on the point that these actions were oriented to influence the public, and through that avenue, influence the actions of the Lufthansa corporation, rather than badgering the airline into conceding to a set of demands. Specifically, the judge ruled that the protest was not an action of force intended to compel an AGAINST THE MAN 141 action from Lufthansa; the action's intention was to impact public opinion first.[57]

Not all jurisdictions take such an approach, for example, the US vehemently goes after DDoS attackers, as can be witnessed through Anonymous' actions against the Church of Scientology.[58]

3.4 *Airports and Public Entities*

On 29 June 2011, approximately 50 flights at Terminal 3 Indira Gandhi International Airports in Delhi, which is India's busiest airport, were delayed. The delays were caused by a Common Use Passengers Processing System (CUPPS) failure which lasted 12 hours. Although the CUPPS handles check-in counters, boarding gates, and information about the arrival and departure times for the terminal, the delays were only minimal amounting to a maximum of 15-25 minutes for some flights because passengers could be manually checked-in. No flights were cancelled. It was initially thought to be a technical failure. However, following investigation by the Central Bureau of Investigation of India, the matter is now being investigated as a malware cyberattack on the CUPPS. The case was registered under the Information Technology (IT) Act of India and the initial investigation provided evidence that malicious code, likely requiring expert knowledge to imple-

54 M. Sauter, *The Coming Swarm DDoS Actions, Hacktivism, and Civil Disobedience on the Internet*, First Edition, (Bloomsbury Academic, 2014), at 140.

55 *OLG Frankfurt am Main*, 1 SS 319/05, 22 May 2006.

56 Thing, 'Higher Regional Court Says Online Demonstration Is Not Force', 2 July 2006, http://www.post.thing.net/node/1370. *See*, Courage, 'German Precedent Upholds Online Civil Disobedience', 22 October 2014, https://couragefound.org/2014/10/german-precedent-upholds-online-civil-disobedience/.

57 M. Sauter, 2014, at 141-142.

58 *See*, Chapter I.4.4.

ment, was found. In addition to looking for external hackers, the investigation revealed "serious security lapses"[59] at the airport and opened up the possibility that officials, directly or indirectly, may be involved. The investigation is still on-going.

An additional consideration, which comes to light here, is that the airport is publicly owned by the Airports Authority of India which is under the Ministry of Civil Aviation. Thus, cyberattacks must not simply be a consideration for private entities. This conclusion is furthered by the numerous other cyberattacks against public airports, as well as those against NAAs – such as the FAA and the Saudi CAA – and State-owned airlines as contained in Table 1.

In August 2012, a Boston digital security firm uncovered malware hidden in the virtual private network (VPN) of a major non-US international airport.

Virtual Private Network: This allows devices to connect over a public network, so data can be sent and received as if the devices were directly connected to a private network. This provides the user with a level of anonymity.

The firm said that the threat could have compromised anything from the employees' personal information, to the safety of passengers. The cyberattack used a Citadel Trojan to read the screens of employees who logged in remotely to the airport's VPN.[60] The identity of the airport was not published. Trusteer, a cybersecurity company, stated:

> This was potentially very dangerous, but we don't know whether the attacker group was targeting the financial system of the airport for economic gain or if the attack was terrorism-related. [...] They could have been trying to access critical infrastructure – possibly air-traffic control systems and even the air-conditioning ducts on planes. Or they might have been looking at the hiring process, to see if they could get someone in there to work as an employee.[61]

The airport VPN was immediately disconnected after officials were made aware of the breach following a routine security check and authorities have been investigating.

59 M. Kakkar, 'CBI Believes Cyber Attack Led to IGI Airport's Technical Problems in June', ZDnet, 25 September 2011, http://www.zdnet.com/article/cbi-believes-cyber-attack-led-to-igi-airports Aviation Stakeholder:Airports-technical-problems-in-june/.

60 E. Iasiello, 'Aviation and Cyberspace: Convergence of Domains, Convergence of Threats', March *Air and Space Power Journal* 3-9 (2013).

61 M. Dolgow, 'Cyberwars Reach a New Frontier: The Airport', 15 August 2012, https://www.bloomberg.com/news/articles/2012-08-15/cyberwars-reach-a-new-frontier-the-airport.

In July 2013, the Turkish authorities revealed that the passport control system at the departure terminal of the Istanbul Ataturk Airport International was subjected to a cyberattack. It is suspected that malware was used by the attackers to steal data, such as passenger passport information. Potential sabotage motives by the attackers were also not ruled out by the Turkish authorities. A further cyberattack was reported by the media against the passport control system at the Sabiha Gokcen International Airport in Istanbul, which crashed due to the malfunction of the Istanbul provincial security directorate's Polnet data system. Authorities eventually restored the systems at both airports following the necessary recovery operation and, while the cyberattack did not have an impact on safety, it had an impact on flight departures. Turkish authorities have yet to identify the source of the cyberattack.[62]

In Summer 2013, Federal authorities in the US alerted the Center for Internet Security (CIS), which is a US non-profit entity that works with the US Government, that four airports had potentially experienced an advanced persistent threat (APT).

Advanced Persistent Threat (APT): A cyberattack that uses multiple clandestine phases to extract information over a long period of time.

Following further investigation, it was found that a total of 75 airports in the US had come under attack. The cyberattack was an APT which, in this case, had gained access to the airports' networks, remaining hidden, and had stolen sensitive information for an extended period of time. During the investigation, it became apparent that the cyberattack was implemented though a phishing scam email to individuals employed in the aviation industry. The Center for Internet Security concluded in its annual report that the cyberattack was carried out by an undisclosed nation state seeking to breach US commercial aviation networks.[63] The source has still not been determined.

3.5 *Unmanned Aircraft*

In addition to manned aviation stakeholders being subjected to cyberattacks, this is also the case within the context of unmanned aviation.

Unmanned Aircraft: An aircraft without a pilot on board that is flown or is designed to be flown automatically, autonomously, by a remote pilot or using a combination

62 M. Kumar, 'Cyber Attack Hits Istanbul Airport Passport Control System', The Hacker News, 26 July 2013, https://www.thehackernews.com/2013/07/Istanbul-airport-cyber-attack-virus.html.

63 W. Welsh, 'Phishing Scam Targeted 75 US Airports', Information Week, 23 June 2014, http://www.informationweek.com/government/cybersecurity/phishing-scam-targeted-75-us-airports Aviation Stakeholder:Airports/d/d-id/1278762.

of the above.

An example of this took place on 4 December 2012 when a US Lockheed Martin RQ-170 State-operated unmanned aircraft was captured in Iran near the Iran-Afghanistan border. The event was broadcast to the public on Iranian television as a display of the technological capabilities of the State. It also took the opportunity to examine the aircraft and its components, including the data acquired during the aircraft's missions. Therefore, such extracted data could provide exact information on the history of the aircraft, and an analysis of the parts could allow for a better understanding of the aircraft and its vulnerabilities.

In October 2011, a computer virus infected the cockpits of US military Predator and Reaper unmanned aircraft, which logged every keystroke the pilots made while the aircraft was flown remotely on missions over warzones, including Afghanistan. The virus was detected by the military's Host-Based Security System. It did not prevent pilots at Creech Air Force Base in Nevada from flying their missions overseas nor were there any confirmed incidents of classified information being lost or sent to an outside source. However, the removal of the virus required multiple efforts indicating that the virus was resilient, so some sources have concluded that it likely originated from an advanced attacker like another State.[64]

The former South Korean Defence Minister, Kim Tae-Young, declared in 2010 that North Korea had imported mobile truck jamming systems that have the capacity to interfere with GPS signals within a 100 km radius.[65] South Korea has since reported on numerous occasions that North Korea has used jamming equipment to interfere with GPS signals around South Korea's two largest airports near Seoul and across the centre of the Korean peninsula. Reports state that in 2016, North Korea broadcast a jamming signal on 100 occasions which affected 962 aircraft.[66] The effect was limited as the older Inertial Navigation System could be used which does not rely on external signals. The reported harmful interference also reportedly resulted in the loss of a South Korean unmanned aircraft within the territory of South Korea.[67]

---

64 N. Shachtman, 2011.

65 Infosec Institute, 'Hacking Drones … Overview of the Main Threats', 4 June 2013, https://resources.infosecinstitute.com/hacking-drones-overview-of-the-main-threats/.

66 K. Mizokami, 'North Korea Is Jamming GPS Signals', Popular Mechanics, 5 April 2016, https://www.popularmechanics.com/military/weapons/a20289/north-korea-jamming-gps-signals/.

67 Y. Henri, 'The ITU: Challenges in the 21st Century', Lecture presented to the Advanced Masters in Air and Space Law at Leiden University, Leiden, The Netherlands, 19 March 2014. *See*, P. Chan-Kyong, 'N. Korea's Electronic Jamming Signals Against S. Korea On Rise', Phys Org, 2 May 2012, https://www.phys.org/news/2012-05-korea-affects-flights-seoul.html.

Not all examples are within the military context, as in July 2012, Todd Humphreys of Cockrell School of Engineering explained, in his Statement on the Vulnerability of Civil Unmanned Aerial Vehicles and Other Systems to Civil GPS Spoofing,[68] the ease in which an unmanned aircraft could be spoofed through GPS. This was established by a demonstration wherein the target unmanned aircraft was remotely tricked, from half-a-mile away into a commanded dive that was only aborted 3 m above the ground in order to prevent it from crashing. This demonstration exposed the potential ease of hacking into an unmanned aircraft that has limited or no cybersecurity installed features, such as unencrypted communication links. While the aircraft avoided destruction, this may not be the case with malicious cyberattacks, which may also target other air space users or people and property on the ground.

3.6 *Passengers*

Passengers may also be victims of cyberattack, such as when on board an aircraft. In February 2016, journalist Steven Petrow reported that a passenger sitting behind him accessed his email account during an internal US flight from Dallas to Raleigh through the in-flight WiFi on an American Airlines aircraft.[69] Petrow only became aware of the matter when the attacker informed him of what happened, confessing also to accessing other passengers' emails. This was purported as possible as American Airlines provides Gogo, an in-flight Internet service, which is public and "operates in the same ways as most open WiFi hotspots on the ground".[70] Thus, the risk posed by public networks on the ground, such as with the free WiFi provided at airports, applies also to on board Internet. This issue was again raised by Ruben Santamarta of IOActive who was able to access passengers' laptops, phones and tablets through the on board WiFi.[71]

3.7 *Other Stakeholders*

While a cyberattack may focus on one aviation stakeholder, such as that targeted towards an airline's ground infrastructure, this may also have an effect on other entities, such as ground handling companies and airports. Thus, more than one stakeholder may be affected by a single cyberattack.

68 T. Humphreys, 'Statement on the Vulnerability of Civil Unmanned Aerial Vehicles and Other Systems to Civil GPS Spoofing', Submitted to the Subcommittee on Oversight, Investigations, and Management of the House Committee on Homeland Security, 18 July 2012.

69 S. Petrow, 'I Got Hacked Mid-Air While Writing an Apple-FBI Story', USA Today, 24 February 2016, http://www.usatoday.com/story/tech/columnist/2016/02/24/got-hacked-my-mac-while-writing-story/80844720/.

70 *Ibid.*

71 *See*, Chapter II.8.6.

This Chapter has focused on the core aviation stakeholders that have been victims, but it is crucial to state that other stakeholders may be affected. This includes those also identified in Table 1, but could also include other existing aviation stakeholders and new stakeholders in the future. Therefore, the scope of victims is not a static term.

3.8 Accidental v. Deliberate

A cyberattack may come from deliberate and directed acts. However, cyber threats may also be accidental, or attackers may utilise unwitting and unwilling individuals. This can be classified as 'human factors' and these must be factored into cybersecurity measures.

Phishing is one cyberattack method that exploits the human factor vulnerability as this requires contacting individuals and enticing them to act in a way that provides the attacker with sought-after information. It is vital to identify this as a weakness as it can be relatively cheap and easy to exploit this vulnerability, and is often quite easy to strengthen via adequate training, ensuring accountability and creating a culture of cybersecurity.

Accidental acts should not be viewed as one type of act. On the other hand, the threat can be an accidental one from those on the inside. An insider could accidently provide hostile actors with direct access to networks for disruption, destruction or manipulation. This could happen via the user introducing a virus to the network, such as the case against the International Space Station (ISS) where an astronaut introduced a virus into the ISS's computer system. However, on the other hand, an individual may be using their own technology, which unintentionally interferes, for example, Bojczak bought a GPS jammer for his car in order to hide from his employer. The Federal Communications Commission fined Bojczak nearly USD 32,000 after it concluded that in 2012 he had "interfered with Newark Liberty International Airport's satellite-based tracking system when he used an illegal GPS jamming".[72]

Natural disasters, fire, pollution, dust, corrosion, weather, wild life, space electromagnetic storms, water, explosions, radiation and climatic conditions could all cause issues for cybersecurity.[73] These would not constitute a cyberattack, but precautions must be taken to reasonably protect against them.

72 S. Strunsky, 'N.J. Man Fined $32K for Illegal GPS Device that Disrupted Newark Airport System', NJ.com, 8 August 2013, http://www.nj.com/news/index.ssf/2013/08/man_fined_32000_for_blocking_newark_airport_tracking_system.html.

73 *See*, European Parliament, 'Cybersecurity in the EU Common Security and Defence Policy (CSDP) Challenges and risks for the EU', EPRS/STOA/SER/16/214N, PE 603.175, May 2017, at 73.

Technical factors are another type of accidental vulnerability. Rather than humans producing the vulnerability, it is the technology itself that creates the vulnerability. The technology can cause technical issues, or can be inadequate and act as an invitation for attackers. Therefore, investment in cyber secure technology is also required.

4 AVIATION CYBER ATTACKERS

4.1 Introductory Remarks

Highlighting the different people and entities that could or do conduct cyberattacks against civil aviation stakeholders is another important exercise. This is because it will determine whether such cyberattacks are being carried out by a small group of individuals, either acting independently or in a more organised framework, so demonstrating the simplicity of cybersecurity or whether the situation is more complex involving numerous people and entities from different backgrounds, with varying degrees of resources and training at their disposal, with a plethora of motives inspiring such cyberattacks.

Attacker: A person or group of persons that deliberately, whether directly or indirectly, carry out a cyberattack.

Therefore, this subchapter will be dedicated to highlighting and discussing the different types of attackers that target aviation stakeholders.

4.2 Categories of Attackers

The cliché of either a spotty teenager living in the parent's basement trying to cause disruptions to 'unethical corporations' or sleek international criminals trying to steal untold amounts of money, which are so often portrayed in movies, fails to encapsulate the versatility of the attacker landscape.

There are three generally accepted types of attackers, which are grouped into various metaphorical coloured hats (black, white and grey):

1. Black Hat: These undertake harmful or illegal activities, such as stealing personal details or money, or denying individuals access to their data.
2. White Hat: These help fight cyberattacks and improve cybersecurity via legal methods, such as removing malware or conducting penetration tests known as PenTests, to discover vulnerabilities. These are sometimes referred to as 'ethical' attackers.

3. Grey Hat: These are in between Black and White Hats, falling into the grey area between legal and illegal acts, but do not have harmful intent.

An addition three metaphorical coloured hats (red, blue and green) are sometimes included:

1. Red Hat: These target Black Hats, even via illegal methods, with the aim of stopping their illegal activities. These are sometimes labelled as 'vigilantes'.
2. Blue Hat: These implement targeted cyberattacks based on revenge.
3. Green Hat: These are malicious attackers with a basic level of skill.

The motive, intention and level of skill is then relevant for assigning the colour of hat. Further, attackers may switch hats between their activities depending on the specifics of each action, so classification is not static.

While the activities of Black, Blue and Green Hats are often given priority in cybersecurity discussions, White Hats should not be overlooked as they can strengthen cybersecurity. The European Organisation for the Safety of Air Navigation (Eurocontrol), for example, has actively utilised White Hats in the past to benefit from their experiences.[74]

Penetration Test or PenTest: A test to assess how a system will react to a cyberattack, to see whether the system's defences can be penetrated and what information can be extracted.

This categorisation system identifies three points:

1. Not every attacker is focused on illegal activities and will, thus, not pose a threat to aviation stakeholders.
2. Not every attacker is an expert, rather there are varying degrees of skills among individuals and, because of this, a Black Hat hacker is more likely to launch a sophisticated cyberattack, which may be more difficult to protect against, as compared to a Green Hat who may use a public domain code, which a basic antivirus software should identify and block.
3. While a cyberattack may appear unexpected and unprovoked, this is not always the case. A victim may not be targeted at random, but because of a motive, such as for financial gain by a Black Hat or revenge by a Blue Hat.

Thus, appreciating this point can enhance predictability, which can enhance cybersecurity.

---

74 Eurocontrol, for example, has utilised White Hats in the past to benefit from their experiences. *See*, Eurocontrol, 'Hackathon', https://hackathon.eurocontrol.int/.

Not only does this categorisation system demonstrate the fragmented landscape of hackers, it also provides a level of predictability. This is because the attacker is a variable and what determines who attacks depends on what the victim has to offer or is doing. A large multinational financial institute, which has vast capital resources, can predict that it is more likely to be targeted by Black Hats who want to steal resources. Blue Hats may not possess the skills to infiltrate a large organisation that invests in its cybersecurity and such an organisation may be too hidden from the general public to warrant acts of revenge. Therefore, it may be more prudent for it to focus its cybersecurity resources on protecting or removing vulnerabilities that could allow the extraction of resources. Whereas, those potential victims performing basic public services (i.e. restaurants, hotels and retail) may wish to strengthen their websites against DDoS cyberattacks as they expose themselves more to the public and, thus, potentially disgruntled customers. Revenge acts by Blue Hats could then follow. This is of course based on assumptions, so cybersecurity measures should be applied in less likely areas, but as an individual's or company's resources are finite, it does allow the available resources to be allocated based on risk.

4.3 *Nation State and State-Sponsored Actors*

A State may openly advertise its involvement in cyber activities, such as within the military context as many have units responsible for cybersecurity and cyberattacks. For example, the German military (*Bundeswehr*) launched a 'cyber army' in April 2017. The Ministry of Defence (*Bundesministerium für Verteidigung*) announced that it "wants to deploy around 13,500 soldiers and civilian workers by 2021 to protect the *Bundeswehr's* networks and weapons systems, but the unit must also be capable of launching their own attacks against hackers".[75] A State may also not take steps to hide a cyberattack and may even advertise it like the Iranian Government did when it caused a US unmanned aircraft to crash in December 2011 via spoofing.

It is often difficult or impossible to attribute a cyberattack to a State or State-sponsored actors when the relevant State wishes to cover it up, but certain attributes of the cyberattack may give clues as to the source. For example, the Stuxnet computer worm was a highly complicated cyberattack that was targeted against the Iranian nuclear programme by collecting information on industrial systems and causing centrifuges to tear themselves apart by overspinning. It is argued by some cyber experts that because of the intended target and the worm's sophistication, such a cyberattack could only result

75 The Local, 'Germany Building New Army of 13,500 'Cyber Soldiers'', 5 April 2017, https://www.thelocal.de/20170405/germany-building-new-army-of-13500-cyber-soldiers.

from a State.[76] While no State has claimed responsibility, and is unlikely to do so, some cyber experts have attributed the cyberattack to the US or Israel.[77] This was the case when Sweden attributed Russia to the cause of its air traffic control system going off-line in November 2015.

A State may also outsource such tasks so as to distance itself because of political and legal considerations. These are then State-sponsored attackers that act in the interest of the backing State. It has been reported by ProofPoint, a US security company, that Chinese State-sponsored actors spy on military and aerospace organisations in Russia and Belarus.[78] China is not the only State reportedly to be engaged in the sponsorship of cyberattacks. Russia is also reportedly involved as, for example, those involved in the December 2016 cyberattack against the financial, transport and energy sectors of Ukraine have been linked to the Russian State.

The cyberattacks against Estonia, which began on 27 April 2007, swamped websites of numerous Estonian entities, including the Estonian Parliament, banks, ministries, newspapers and broadcasters. The attack methods ranged from DDoS cyberattacks, spamming larger news portal commentaries and defacing websites, such as the Estonian Reform Party website. During this time, the Estonian Foreign Minister was quick to accuse Russia stating that it was trying to paralyse Estonia's economy and Government. The Russian Government denied carrying out the cyberattack, with the then Russian Parliamentary leader Sergei Markov suggesting Estonia look elsewhere than the Russian Government. This assertion was based on additional knowledge, as Markov's assistant, a leader in a pro-Putin movement called 'Nashi', had openly admitted to the cyberattack. This type of attack has been labelled as 'patriotic hacking' as it is not directly linked to the State, but it shares similar or the same goals and ideologies with that State.[79]

These examples show the complexity as State attribution may take several forms, as:
1. a State may directly carry out the cyberattack,
2. the cyberattack may be sponsored by the State,

---

76 J. Ohlin, K. Govern and C. Finkelstein (eds), *Cyberwar: Law and Ethics for Virtual Conflicts*, First Edition, (Oxford University Press, 2015), at 13.

77 E. Nakashima and J. Warrick, 'Stuxnet Was Work of US and Israeli Experts, Officials Say', Washington Post, 2 June 2012, https://www.washingtonpost.com/world/national-security/stuxnet-was-work-of-us-and-israeli-experts-officials-say/2012/06/01/gJQAlnEy6U_story.html?utm_term=.75886d813934.

78 P. Paganini, 'Chinese State-Sponsored Hackers Targets Russia and Belarus with ZeroT and PlugX', Security Affairs, 3 February 2017, http://www.securityaffairs.co/wordpress/55942/apt/chinese-state-sponsored-aerospace.html.

79 P. Singer and Allan Friedman, *Cybersecurity* and *Cyberwar: What Everyone Needs to Know*, First Edition, (Oxford University Press, 2014), at 110.

3. the State may produce a rhetoric or provide a legal system that does not adequately address cyberattacks so to induce cyberattacks, or
4. a cyberattack may have a desirable outcome for the State, but that State has no direct or indirect involvement.[80]

As States are unlikely to claim responsibility or identify a link to an attacker, it may never be verified which of the four levels of State involvement exists. This is evident in Table 1 as most of the cyberattacks linked to States are unconfirmed, with potentially more examples involving States.

The inclusion of States as actors, thus shows the relevance of national security for aviation cybersecurity. Regulators must clarify the degree of interaction that aviation stakeholders have with national security, which is typically a function of the State. Thus, national security, aviation security and aviation safety are three areas of consideration.

4.4 *Hacktivists and Hacktivism*

The terms 'hacktivists' and 'hacktivism' were coined in 1994 by a Cult of the Dead Cow member known as 'Omega' in an email to other group members.[81]

Hacktivism: This is a politically or ideologically motivated cyberattack. This involves hostile and sometimes destructive cyberattacks in support of a cause or belief.

Hacktivism is broader than 'patriotic hacking' as it does not have to be linked to a State. However, the exact scope of both 'hacktivism' and 'hacktivists' is ambiguous as they are centred on subjective political or ideological beliefs. As a result of this, they are fluid terms.

Members of groups like Anonymous, for example, are often cited as being hacktivists as they engage in operations against targets to punish a perceived transgression or draw attention to a situation. A famous Anonymous case of hacktivism involved the group's conflict with the Church of Scientology. In 2008, members of Anonymous launched a DDoS cyberattack to disrupt the victim's webpage. Such acts were often traceable, as it involved Script Kiddies[82] using off-the-shelf software, and this resulted in some of the

80 *See*, D. Pinchuk, 'Patriotic Russians May Have Staged Cyber Attacks on Own Initiative: Putin', Reuters, 1 June 2017, http://www.reuters.com/article/us-russia-economic-forum-putin-cyber/patriotic-russians-may-have-staged-cyber-attacks-on-own-initiative-putin-idUSKBN18S56Y.

81 O. Ruffin, 'Hacktivism, from Here to There', Cult of the Dead Cow, http://www.cultdeadcow.com/cDc_files/cDc-0384.php.

82 *See*, Chapter I.4.8.

participants being prosecuted. Dmitriy Guzner, a 19-year-old US citizen, pleaded guilty in May 2009 for using a botnet to attack the Church of Scientology's website with false traffic in January 2008. Guzner was sentenced to a year and a day in prison, a two-year probation period following the prison sentence and ordered to pay USD 37,500 in restitution.[83]

The DDoS cyberattack against Lufthansa in June 2001 may also be defined as hacktivism as its intention was to draw attention to the airline's role in exporting asylum seekers on behalf of the German Government.

4.5 Journalists

In August 2016, it was reported that Russian intelligence had hacked media sources in the US. In the case of the New York Times, reporters had been targeted by spear-phishing cyberattacks. It is not surprising that journalists and their affiliated companies are attacked as they are often involved in sensitive investigative journalism which could embarrass States or ruin the reputation of companies. However, their involvement in cyberattacks goes both ways.

A recent and significant example of journalist involvement in a cyberattack was the 'Panama Papers' incident, which involved 11.5 million leaked documents that detail financial and attorney-client information of more than 214,488 off-shore entities that were held by the Panamanian law firm and corporate service provider Mossack Fonseca. Some of the information belonged to wealthy individuals and public officials. While off-shore business entities are legal, reporters found that some of the Mossack Fonseca shell corporations were used for illegal purposes, such as tax evasion and evading international sanctions. The documents were then leaked by the whistle-blower 'John Doe' to the German journalist Bastian Obermayer who was working for the newspaper *Süddeutsche Zeitung* and were later published, upon verification, by the International Con-

83 *United States v. Guzner*, Case 2:08-cr-01224-UA (2008). The indictment stated 18 USC § 1030(a)(5)(A) and (B): Unauthorized Impairment of a Protected Computer.

(a)Whoever-

[...]

(5)

(A) knowingly causes the transmission of a program, information, code, or command, and as a result of such conduct, intentionally causes damage without authorization, to a protected computer;

(B) intentionally accesses a protected computer without authorization, and as a result of such conduct, recklessly causes damage;

[...]

shall be punished as provided in subsection (c) of this section.

sortium of Investigative Journalists.[84] This shows the active role that journalists play in this area.

Journalist involvement in cyberattacks has also taking a much more controversial role. For example, the News International phone-hacking scandal involved the defunct News of the World newspaper, as well as other British newspapers published by News International. Employees of the newspaper were accused of engaging in phone hacking, police bribery and exercising improper influence in the pursuit of stories. While investigations conducted from 2005 to 2007 appeared to show that the newspaper's phone-hacking activities were limited to celebrities, politicians and members of the British Royal Family, in July 2011 it was revealed that the phones of murdered schoolgirl Milly Dowler, relatives of deceased British soldiers and victims of the 7 July 2005 London bombings had also been targeted. The former UK Prime Minister, David Cameron, announced on 6 July 2011 that a public inquiry, known as the *Leveson Inquiry*, would look into this matter.[85] As of yet, no aviation example could be found.

4.6 *Terrorists and Criminals*

Black and Green Hats deserve a special mention as these could incorporate those with minimal skills that simply seize an easy opportunity to exploit a vulnerability, to skilled professionals, some of whom rely on cyberattacks as their primary occupation or terrorist groups.

While terrorists and terrorist organisations have preferred physical attacks in the past, such as suicide bombing, there is a growing threat within the cyber domain. Terrorist organisations have used technology to recruit, disseminate propaganda, incite, radicalise, finance, train, plan and research. However, there are cases where terrorist leaders have encouraged members of their organisation to utilise the Internet for more operational purposes. For example, Imam Samudra, the individual responsible for engineering the 2002 Bali nightclub bombings, published a document in 2004 with information on using cyberattacks against US interests with the aim of bankrupting the State.

Other actors that are also involved in cyberattacks, but do not come under the heading of terrorists, are criminals. The definition of 'criminal' will depend on the jurisdiction as not all States have enacted cybercrime legislation and others do not have mature cybercrime

84 International Consortium of Investigative Journalists, 'Panama Papers', https://www.icij.org/tags/panama-papers/.

85 *Leveson Inquiry*, 'An Inquiry into the Culture, Practices and Ethics of the Press Report', The Right Honourable Lord Justice Leveson, Vol. 1-4, November 2012.

rules to sufficiently encompass all cyberattacks. Therefore, some cyberattacks would be a cybercrime in one jurisdiction, but not another. Also, there is a clear overlap between criminal and terrorist activities. Criminal acts can be conducted either by individuals acting alone or within a more organised structure involving numerous people. Such acts are often motivated by money or gaining information. For example, it was reported in 2013 that the cryptolocker ransomware made over USD 30 million in 100 days. Ransomware, which is a common type of cyberattack, blocks the use of a computer and requires the victim to pay a sum of money to get the computer unlocked, like that used against DLA Piper, a law firm with an aviation practice, and the NHS in 2017. Other common methods include taking control of the victim's computer, extracting information and selling it to a third party.

4.7 Insiders and Opportunists

The broad and unpredictable scope of potential attackers is furthered by the inclusion of insiders and opportunists. Insiders can take the form of an employee that, for example, is unhappy with their employer because the employee has been fired or disagrees with the work that their employer is conducting. The cyberattack could, for example, be against the company's network, such as the Omega Engineering Case, where a disgruntled employee, Timothy Lloyd, sabotaged the company's network with a logic bomb in 1996.[86] He apparently did this as an act of revenge for being fired. The attach cost the company USD 12 million in network damages and forced the eventual layoff of around 80 employees.

Another way that insiders may take revenge on a company is to transmit proprietary information to competitors, or release confidential or sensitive information to the public. An example of this was the US soldier, Chelsea Elizabeth Manning case, who was convicted by court martial in July 2013 under the Espionage Act and other offences after disclosing to WikiLeaks nearly three-quarters of a million classified or unclassified sensitive, military and diplomatic documents.[87]

Rather than an employee, the threat may come from a contractor, like the case of Edward Snowden who was a Central Intelligence Agency (CIA) employee and contractor for the US Army who copied and leaked classified information from the US National Security Agency (NSA) in 2013 without authorisation.

86 *United States v. Timothy Lloyd*, No. 00-2409 (2001).

87 *United States v. Manning*, Court Martial, US Army Trial Judiciary, 1st Judicial Circuit (2013).

4.8 Anyone and Script Kiddies

It is not a new phenomenon for members of the public who have no affiliation to criminal organisations or terrorist groups to engage in criminal activities. Within the context of international air law, this can be demonstrated by the Tokyo Convention 1963, which regulates the behaviour of passengers.[88] Therefore, it should also not be dismissed that the scope of attackers could include anyone.

The inclusion of anyone can be best demonstrated via the rise of script kiddies.

Script Kiddie: A person who uses existing computer scripts or codes to implement a cyberattack due to a lack of expertise preventing them from writing their own.

There is a misconception that script kiddies are naive and immature juveniles giving the term derogatory connotations. However, it does not relate to the age of the participant or their understanding of their actions, rather it pertains to individuals possessing a limited set of cyber-based competencies that often rely on external sources, such as utilising codes written by others with more experience, to launch a cyberattack against the victim.

There are specific websites providing downloadable programmes that are capable of breaching computers and networks. These are easily available where the script kiddie can download programmes and then send to a third party. The ease of downloading sophisticated programmes capable of causing serious effects is part of the rise of script kiddies as it gives a powerful tool to an untrained demographic.

A cyberattack could also be initiated by someone with no specific skills in this domain. In February 2016, a Twitter account named *@GunnexGod* wrote: "Hello @Ryanair, you have 15 minutes before I commit the biggest terror attack the UK has ever seen on one of your planes. Be ready". Ryanair took the threat seriously and managed to trace the tweet to Brian Lake from Pennsylvania, US. The airline filed extortion and defamation proceedings claiming the tweet had "specific intent of harming Ryanair's business and its reputation of providing high-quality, reliable, low-cost goods and services in the commercial airline industry".[89] A US District Court judge ordered Brian Lake to pay a total of USD 284,148.[90] Anyone with a computer and Internet access can create a Twitter account and

---

88 *See*, Chapter III.7.2.1. Convention on Offences and Certain Other Acts Committed on Board Aircraft (Tokyo Convention), 704 UNTS 219; 20 UST 2941; 2 ILM 1042, 14 September 1963, Art. 1(1).

89 S. Calder, 'Ryanair Wins £220,000 from Twitter User Who Posted Terror Threat', The Independent, 14 July 2017, http://www.independent.co.uk/travel/news-and-advice/ryanair-bomb-threat-terror-twitter-brian-lake-pennsylvania-extortion-a7841446.html.

90 *Ryanair v. Brian Lake*, 2:17-cv-00166 (2017).

cause disruption to international transport, so demonstrating the possibility that anyone could engage in cyberattacks.[91]

What is potentially more worrying for cybersecurity is that anyone with a computer can be part of a cyberattack without knowing or intending to be a part. A famous example of this is the Anna Kournikova Virus. This was a worm released in February 2001 that spread by email, disguised as an email attachment with the file name 'AnnaKournikova.-jpg.vbs'. The virus was a script that, once opened, emailed itself to all the contacts within the user's address book. The virus spread very quickly but was fortunately a payload-free virus and did not erase any files or send information back to the creator.

> Payload: This refers to the part of a computer virus that executes the malicious activity.
>
> Payload Free: This is a virus that does not have a code that executes a malicious activity.

Dutch programmer Jan De Wit turned himself in to the authorities when it became clear how wide the virus had spread. De Wit was charged with spreading data into a computer network with the intention of causing damage, a crime that carried a maximum sentence of four years in prison and a fine of circa USD 41,300.[92] De Wit was sentenced to 150 hours community service or 75 days in jail due to the limited damage caused by the virus. This example shows how the acts of unsuspecting people, i.e. the receivers of the emails, contribute to the proliferation of the malware. This is because if they had not opened the email and downloaded the attachment, then that computer would not have become infected and the worm would not have been able to access the user's address book and spread further. Therefore, this shows that anyone can be part of a cyberattack.

4.9 Attribution

Traditional security attacks on aviation, such as hijacking and airport bombings, can usually be attributed to individuals or organisations. This is because the attacker must be present at the scene, at least for a short period of time, and may also claim responsibility afterwards due to the often-political intentions motivating the attack. This was seen, for example, with the tragic 22 March 2016 bombing at Zaventem Airport in Brussels,

91 Such incident, due to its low skill requirement, may not constitute a cyber-attack, but rather a criminal act that utilised a computer.

92 J. Leyden, 'Anna Kournikova Virus Author Stands Trial: Lenient Sentence in Prospect', The Register, 14 September 2001, https://www.theregister.co.uk/2001/09/14/anna_kournikova_virus_author_stands/.

Belgium. The individual bombers were identified and Islamic State of Iraq and the Levant (ISIL) subsequently claimed responsibility for the attack. However, as highlighted in Table 1, not every cyberattack is so easily attributable with many going unclaimed.

Attribution, in the context of a cyberattack, can take one of three forms:

1. Unclaimed or unattributed: Neither the attacker identifies themselves nor can investigators, whether by law enforcement or other means, attribute the cyberattack.[93] An example of this is the LOT airlines incident in June 2015.
2. Claimed or attributable, but the personal identity of the involved individuals is unknown: A cyberattack may be claimed by or attributed to an entity, such as Anonymous, or an attacker via their pseudonym. However, the personal identity of the attacker remains unknown. An example of this is the Mallu Cyber Soldiers, who claimed responsibility for the December 2016 cyberattack against Sialkot International Airport.
3. Claimed or attributable, and the personal identity of the involved individuals is known: It may be the case that the attacker wants to be known or the attacker's identity is discovered following investigation. An example of this is Paul Dixon who orchestrated a cyberattack against British Airways' website in October 2014 and who was identified by law enforcement.

As highlighted above, an additional point to attribution is law enforcement. From the three cases listed above, only Paul Dixon was arrested and appeared before a court. He was given a suspended two-year jail sentence and ordered to pay GBP 200 after admitting to five counts of unauthorised modification of computer material following a DDoS cyberattack against British Airways and several police forces in the UK.[94] It may be the case that an attacker surrenders themselves to law enforcement, as De Wit did following his release of the Anna Kournikova Virus in February 2001.

While a cyberattack may be attributable to an individual or group, this does not mean that law enforcement can apprehend and enforce the law of the affected State. This is because a cyberattack has the potential to be launched anywhere in the world, so the attacker may be in another jurisdiction. Therefore, other areas of law, such as jurisdiction, extradition and immunities[95] – as presented in the Edward Snowden case highlight – will

93 This results in the attacker remaining anonymous; the true scope of the cyberattack being difficult to ascertain; and recovery of any stolen, corrupted or lost data, information or physical property being unlikely.

94 *R v. Paul Dixon*, Newcastle Crown Court (Unreported) 3 April 2017. *See*, for an overview of UK court cases involving the Computer Misuse Act 1990, M. Turner, 'Computer Misuse Act 1990 Cases', http://www.computerevidence.co.uk/Cases/CMA.htm.

95 *See*, for aviation specific cases of jurisdiction and extradition, Chapter III.7. It is, however, outside the scope of this work to look into the general rules of jurisdiction, extradition and immunities.

also be relevant. Cybersecurity should not be viewed in a regulatory silo, as wider issues of jurisdiction, extradition and immunities, such as when States are involved, are also necessary to consider.

II European Union Regulation of Aviation Cybersecurity

1 Introduction

1.1 European Union Cybersecurity Strategy

Prior to 2013, efforts by the Commission and individual Member States to deal with the growing number of cyberattacks were fragmented. As a result, on the 7 February 2013, the EU published a cybersecurity strategy to bring about harmonised efforts to tackle the growing and evolving cyber threat landscape.[1] The cybersecurity strategy – Cybersecurity Strategy of the European Union: An Open, Safe and Secure Cyberspace – presented by the Commission and the High Representative of the Union for Foreign Affairs and Security Policy (High Representative), "outlines the EU's vision in this domain, clarifies roles and responsibilities and sets out the actions required based on strong and effective protection and promotion of citizens' rights to make the EU's online environment the safest in the world".[2]

The 2013 cybersecurity strategy clarified the guiding principles that the EU cybersecurity policy should adopt.

1. Applicability of EU's Core Values: Cybersecurity should be a part of the EU system and not something additional.
2. Protect Fundamental Rights: Cybersecurity should be based on the fundamental rights and freedoms contained in the Charter of Fundamental Rights[3] of the EU and its EU core values (e.g. freedom of expression, personal data and privacy).
3. Access for All: Every EU citizen should have access to the Internet and the unhindered flow of information.
4. Democratic and Efficient Multi-Stakeholder Governance: All stakeholders are important for Internet governance as there is no one stakeholder in control.

---

1 Commission, 'Cybersecurity Strategy of the European Union: An Open, Safe and Secure Cyberspace', Joint Communication to the European Parliament, the Council, the European Economic and Social Committee and the Committee of the Regions, Brussels, Belgium, JOIN (2013) 1 Final, 7 February 2013.

2 Commission, 'Communication on a Cybersecurity Strategy of the European Union – An Open, Safe and Secure Cyberspace', 7 February 2013, https://www.ec.europa.eu/digital-single-market/en/news/communication-cybersecurity-strategy-european-union-%E2%80%93-open-safe-and-secure-cyberspace.

3 Charter of Fundamental Rights of the European Union, 26 October 2012, 2012/C 326/02 (2012).

5. Shared Responsibility: All stakeholders (i.e. public authorities, private entities or individual citizens) have a responsibility for cybersecurity.[4]

The EU's aim was to actualise its EU cybersecurity strategy through five strategic priorities.

1. Achieve cyber resilience.
2. Drastically reduce cybercrime.
3. Develop a cyber defence policy and capabilities related to the Common Security and Defence Policy.
4. Develop industrial and technological resources for cybersecurity.
5. Establish a coherent international cyberspace policy for the EU and promote its core values.

The EU has made significant progress under the 2013 cybersecurity strategy. In doing so, the EU has created a regulatory and institutional environment to increase cybersecurity. For example, the EU has taken regulatory steps to protect Network and Information Systems (NIS Directive[5]) and data (General Data Protection Regulation[6] (GDPR)), which both are important for aviation stakeholders. Further, the EU has set out a robust foundation for the fight against cybercrime[7] in line with the Budapest Convention[8] and has created the European Cybercrime Centre (EC3) at EUROPOL to advance cooperation between national law enforcement agencies.

1.2 *Digital Single Market Strategy for Europe*

Separate, but connected to the cybersecurity strategy is the Digital Single Market Strategy for Europe.[9] The Commission has defined 'digital single market' as a market:

4 Commission, JOIN (2013) 1 Final, 7 February 2013, at 3-4.

5 Directive No. 2016/1148 of the European Parliament and of the Council of 6 July 2016 Concerning Measures for a High Common Level of Security of Network and Information Systems Across the Union, OJ L 194, 19 July 2016, pp. 1-30.

6 Regulation (EU) No. 2016/679 of the European Parliament and of the Council of 27 April 2016 on the Protection of Natural Persons with Regard to the Processing of Personal Data and on the Free Movement of Such Data, and Repealing Directive No. 95/46/EC (General Data Protection Regulation), OJ L 119, 4 May 2016, pp. 1-88.

7 Directive No. 2013/40/EU of the European Parliament and of the Council of 12 August 2013 on Attacks Against Information Systems and Replacing Council Framework Decision No. 2005/222/JHA, OJ L 218, 14 August 2013, pp. 8-14.

8 Convention on Cybercrime, European Treaty Series (Budapest Convention 2001) – No. 185 23.XI.2001, 23 November 2001.

9 Commission, Communication from the Commission to the European Parliament, the Council, the European Economic and Social Committee and the Committee of the Regions: A Digital Single Market Strategy for Europe, Brussels, 6 May 2015, COM (2015) 192 Final, at 3.

> in which the free movement of goods, persons, services and capital is ensured and where individuals and businesses can seamlessly access and exercise online activities under conditions of fair competition, and a high level of consumer and personal data protection, irrespective of their nationality or place of residence.[10]

The EU noted that legal fragmentation and trade barriers, that do not exist in the physical single market, exist in the digital market and these are hindering trade and investment. It has been predicted by the EU that the removal of these barriers within Europe could contribute an additional EUR 415 billion to the combined EU Member States' GDP.[11] This is because the "digital economy can expand markets and foster better services at better prices, offer more choice and create new sources of employment".[12]

In order to achieve a digital single market within the EU, the Commission published its digital single market strategy on 6 May 2015[13] and is built upon three pillars.
1. Provide better access to online goods and services for consumers and businesses.
2. Create the right conditions for digital networks and services to flourish.
3. Maximise the growth potential of the European digital economy.[14]

The digital single market strategy then set out 16 key actions under the three pillars, which the Commission delivered at the end of 2016.[15] Notable areas of work involved the:
1. end of roaming charges,[16]
2. modernisation of data protection,[17]
3. cross-border portability of online content,[18] and
4. agreement to unlock e-commerce by stopping unjustified geo-blocking.[19]

10 *Ibid.*
11 *Ibid.*
12 *Ibid.*
13 *See*, for the text of the strategy, *Ibid.*
14 *Ibid.*, at 4.
15 Commission, 'A Digital Single Market for Europe: Commission Sets Out 16 Initiatives to Make It Happen', Press Release, Brussels, 6 May 2015, http://www.europa.eu/rapid/press-release_IP-15-4919_en.htm.
16 Regulation (EU) No. 2017/920 of the European Parliament and of the Council of 17 May 2017 Amending Regulation (EU) No. 531/2012 as Regards Rules for Wholesale Roaming Markets, OJ L 147, 9 July 2017, pp. 1-8.
17 Regulation (EU) No. 2016/679.
18 Regulation (EU) No. 2017/1128 of the European Parliament and of the Council of 14 June 2017 on Cross-Border Portability of Online Content Services in the Internal Market, OJ L 168, 30 June 2017, pp. 1-11.
19 Regulation (EU) No. 2018/302 of the European Parliament and of the Council of 28 February 2018 on Addressing Unjustified Geo-Blocking and Other Forms of Discrimination Based on Customers' Nationality, Place of Residence or Place of Establishment Within the Internal Market and Amending Regulations (EC) No. 2006/2004 and (EU) No. 2017/2394 and Directive No. 2009/22/EC, OJ L 60I, 2 March 2018, pp. 1-15.

On 10 May 2017, the Commission published a mid-term review of the digital single market strategy, which presents and evaluates the progress in implementing the strategy since 2015 and highlights where further actions are needed.[20] The review identified three main areas where further action is required.

1. Develop the European data economy to its full potential.
2. Protect Europe's assets by tackling cybersecurity challenges.
3. Promote online platforms as responsible players of a fair internet ecosystem.[21]

Therefore, the work is not complete as, for example, the review explores several areas that should be focused on in the future: Digital skills, digitising industry and services (e.g. connected cars), high performance computing, artificial intelligence or AI, modernising public services and e-government, and health and care.

Since May 2015, the Commission has delivered around 35 legislative proposals and policy initiatives as announced in its digital single market strategy.[22] A part of this is the NIS Directive and the EU's push for its full and correct implementation.

1.3 Resilience, Deterrence and Defence in the European Union

In September 2017, the Commission revisited its cybersecurity strategy and updated its focus on 'resilience, deterrence and defence'.[23] In addition to updating the 2013 cybersecurity strategy, it "can also be understood as a stepping stone in the Commission's goal to create a so-called digital single market".[24] The proposal sets out new initiatives.

1. Build a stronger EU cybersecurity agency.
2. Introduce an EU-wide cybersecurity certification scheme.
3. Swift implementation of the NIS directive.

20 Commission, 'Shaping the Digital Single Market', https://www.ec.europa.eu/digital-single-market/en/policies/shaping-digital-single-market.

21 Commission, 'Digital Single Market: Commission Calls for Swift Adoption of Key Proposals and Maps out Challenges Ahead', Press Release, Brussels, 10 May 2017, http://www.europa.eu/rapid/press-release_IP-17-1232_en.htm.

22 Commission, Implementation of the Digital Single Market Strategy to the Communication from the Commission to the European Parliament, the Council, the European Economic and Social Committee and the Committee of the Regions on the Mid-Term Review on the Implementation of the Digital Single Market Strategy: A Connected Digital Single Market for All, Brussels, 10 May 2017, COM (2017) 228 Final, Ann. 1. *See*, European Council and Council of the European Union, 'Digital Single Market for Europe', http://www.consilium.europa.eu/en/policies/digital-single-market/.

23 Commission, Joint Communication to the European Parliament and the Council, Resilience, Deterrence and Defence: Building Strong Cybersecurity for the EU, Brussels, 13 September 2017 JOIN (2017) 450 Final.

24 Global Risk Insights, 'New EU Cyber Strategy Leaves Key Security Gaps', https://www.globalriskinsights.com/2017/10/new-eu-cyber-strategy-leaves-key-security-gaps/.

EU leaders regard cybersecurity reform as one of the main on-going aspects on the road to completing the EU digital single market.[25] Thus, it builds on the measures already put in place by the cybersecurity strategy and digital single market strategy.

In addition to building on on-going activities, it also suggests new initiatives to further improve EU cyber resilience and response.

1. Build EU resilience to cyber-attacks and step up the EU's cybersecurity capacities.
2. Create an effective criminal law response.
3. Strengthen global stability through international cooperation.[26]

The Commission, in its new strategy is, therefore, proposing to reinforce the EU's resilience, deterrence and response to cyber-attacks.[27]

The strategy does not refer directly to the aviation industry, however, it does acknowledge the importance that information and communications technology has for certain sectors, such as finance, health, energy and transport.[28] Aviation is included as part of transport under EU law, as demonstrated by Article 100(2) of the Treaty on the Functioning of the European Union (TFEU)[29] and confirmed by Annex II of Directive (EU) 2016/1148,[30] more commonly known as the NIS Directive,[31] a Directive born out of the strategy, which includes aviation within the scope of 'transport'.

The references made to transport give a clear indication that aviation will be a subject of the EU's cybersecurity efforts, both directly via sector-specific measures and generally through non-aviation rules, such as the NIS Directive. These measures will be the focus of Chapter II. This chapter will be organised around the EU's aviation safety regime as this mirrors the EU's methodology of regulating aviation cybersecurity with other areas of law also being investigated when applicable.

---

25 European Council and Council of the European Union, 'Reform of Cyber Security in Europe', http://www.consilium.europa.eu/en/policies/cyber-security/.

26 Commission, 'Cybersecurity', Fact Sheet, 2017, http://www.consilium.europa.eu/media/21480/cybersecurity factsheet.pdf.

27 *Ibid.*

28 Commission, JOIN (2013) 1 Final, 7 February 2013, at 2.

29 Consolidated Version of the Treaty on the Functioning of the European Union, 2008/C 115/01, 13 December 2007, Art. 100(2). *See*, Chapter II.2.1.

30 Directive (EU) No. 2016/1148 of the European Parliament and of the Council of 6 July 2016 Concerning Measures for a High Common Level of Security of Network and Information Systems Across the Union, OJ L 194, 19 July 2016, pp. 1-30, Ann. II.

31 *See*, Chapter II.8.4.

1.4 *Endorsement of the Strategies: Tallinn Digital Summit*

Endorsement for the Cybersecurity Strategy and Digital Single Market Strategy has been widely achieved. This is evident from the 29 September 2017 Tallinn Digital Summit, which was organised by the Estonian Presidency of the Council of the European Union, in cooperation with the President of the European Council and the European Commission, which endorsed both Strategies. The summit brought together the heads of the 28 EU Member State to discuss "further plans for digital innovation with the aim of keeping Europe ahead of the technological curve while becoming a digital leader, globally, in the years to come".[32] In addition to endorsement, the summit provides a good indication of the opinions of key political and legal players in the EU system, as well as where new laws and policy are going to be created.

1. Create a Digital Single Market.
2. Fair taxation of the digital economy.
3. Build strong cybersecurity through resilience, deterrence and defence.
4. Digital Skills in Europe.
5. Complete the Digital Single Market State of Play.
6. Roadmap for a more united, stronger and more democratic union.
7. Enhance cooperation.
8. EU trust funds to address the root causes of migration.
9. Commission initiatives end 2017-18.
10. State of the Union Address 2017.[33]

Therefore, these areas will receive political and legal attention from the EU and stakeholders.

2 Role of the European Union in Air Transport

2.1 *Legal Basis for Air Transport in the European Union*

The starting point for the EU's competency in aviation is to be found in Article 4 of the TFEU.

32 Estonian Presidency of the Council of the European Union, 'Tallinn Digital Summit', https://www.eu2017.ee/political-meetings/tallinn-digital-summit.

33 Commission, 'Tallinn Digital Summit – Factsheets', https://www.ec.europa.eu/commission/publications/tallinn-digital-summit-factsheets_en.

> The Union shall share competence with the Member States where the Treaties confer on it a competence which does not relate to the areas referred to in Articles 3 and 6.

Article 4(2)(g) of the TFEU then states that such shared competencies include 'transport'. Therefore, in the field of air transport, the EU Member States can exercise their own competence in areas where the EU has not or does not exercise its own competence, as opposed to exclusive competences under Article 3 TFEU.[34]

Article 100(2) of the TFEU then expands upon the measures open to the EU within the scope of air transport.

> The European Parliament and the Council, acting in accordance with the ordinary legislative procedure, may lay down appropriate provisions for sea and air transport. They shall act after consulting the Economic and Social Committee and the Committee of the Regions.

As aviation transport is a shared competence, Member States have the capacity to develop national rules for aviation safety where EU law is silent on the matter.[35]

It must be highlighted that 100(2) is the legal basis for aviation transport, and because of this, other legal basis may be relevant for cybersecurity depending on the specific context. For example, the Cybersecurity Strategy notes that "a particularly serious cyber incident or attack could constitute sufficient ground for a Member State to invoke the EU Solidarity Clause".[36] The Solidarity Clause is in Article 222 of the Treaty on the Functioning of the European Union:

34 EU, 'Division of Competences within the European Union', 26 January 2016, http://www.eur-lex.europa.eu/legal-content/EN/TXT/?uri=LEGISSUM%3Aai0020.

35 *See*, for an example of national aviation cybersecurity rules, UK Department for Transport, Aviation Cyber Security Strategy: Moving Britain Ahead, https://assets.publishing.service.gov.uk/government/uploads/system/uploads/attachment_data/file/726561/aviation-cyber-security-strategy.pdf. The UK Aviation Cyber Security Strategy is an attempt to align the UK Government's National Cyber Security Strategy and the Department for Transport's 2050 Aviation Strategy.

36 Commission, Cybersecurity Strategy of the European Union: An Open, Safe and Secure Cyberspace, 7 February 2013, JOIN (2013) 1 Final, at 19.

1. The Union and its Member States shall act jointly in a spirit of solidarity if a Member State is the object of a terrorist attack or the victim of a natural or man-made disaster. The Union shall mobilise all the instruments at its disposal, including the military resources made available by the Member States, to:
   a.
      - prevent the terrorist threat in the territory of the Member States;
      - protect democratic institutions and the civilian population from any terrorist attack;
      - assist a Member State in its territory, at the request of its political authorities, in the event of a terrorist attack;
   b. assist a Member State in its territory, at the request of its political authorities, in the event of a natural or man-made disaster.

Therefore, the EU and its Member States shall cooperate when a Member State is the object of a terrorist attack or the victim of a natural or man-made disaster. As a result, the specific legal basis must be analysed for each piece of EU secondary legislation, but Article 100(2) is the fundamental reference for aviation.

2.2 *Regulating Aviation Safety in the European Union*

2.2.1 Regulation 216/2008 and the European Aviation Safety Agency

Having regard to Article 100(2), Regulation 1592/2002 of the European Parliament and of the Council, which established the common rules in the field of civil aviation and EASA, was published in 2002.[37] Regulation 1592/2002 was then repealed and replaced in 2008 by Regulation 216/2008 of the European Parliament and of the Council.[38] The change updated the rules in the light of the technical advancements that took place within the aviation industry and the lessons learnt by the Agency through its on-going and numerous activities. Regulation 216/2008 was the legal basis in the EU aviation safety regime and it was this regime, which was at the centre of the EU's cybersecurity aviation transport effort at its inception. It has since been repealed and replaced.[39]

---

37 Regulation (EC) No. 1592/2002 of the European Parliament and of the Council of 15 July 2002 on Common Rules in the Field of Civil Aviation and Establishing a European Aviation Safety Agency, OJ L 240/1, 7 September 2002, pp. 1-21.

38 Regulation (EC) No. 216/2008 of the European Parliament and of the Council of 20 February 2008 on Common Rules in the Field of Civil Aviation and Establishing a European Aviation Safety Agency, and Repealing Council Directive No. 91/670/EEC, Regulation (EC) No. 1592/2002 and Directive No. 2004/36/EC, OJ L 79, 19 March 2008, pp. 1-49. Regulation 216/2008 has been amended four times and an unofficial consolidated version is available at EASA, 'Regulations', https://www.easa.europa.eu/regulations. (Hereinafter: 'Regulation 216/2008'.)

39 *See*, Chapter II.9.

2.2.2 Joint Aviation Authorities

The Joint Aviation Authorities (JAA) was an associated body of European Civil Aviation Conference (ECAC) representing the civil aviation regulatory authorities of a number of European States who had agreed to cooperate in developing and implementing common safety regulatory standards and procedures. The adoption of the Regulation 1592/2002 by the European Parliament and the Council of the EU and the subsequent establishment of EASA created a Europe-wide regulatory authority which has absorbed most functions of the JAA (in EASA Members States). The JAA entered into a new phase as of 1 January 2007. In this new phase, it became the JAA Transition (JAA T) which consisted of a Liaison Office (JAA LO) and a Training Office (JAA TO). In 2009, JAA was disbanded. Only JAA-TO remains and this has not yet included cybersecurity within its portfolio.

2.2.3 Objective and Scope of the European Aviation Safety Agency

Pursuant to Article 2(1) of Regulation 216/2008, the Agency's main objective was to "establish and maintain a high uniform level of civil aviation safety in Europe" within the 28+4 EU Member States.[40] In addition to this, Article 2(2) then set forth six additional objectives.

Additional objectives were, in the fields covered by this Regulation, as followed:

a. to ensure a high uniform level of environmental protection;
b. to facilitate the free movement of goods, persons and services;
c. to promote cost-efficiency in the regulatory and certification processes and to avoid duplication at national and European level;
d. to assist Member States in fulfilling their obligations under the Chicago Convention, by providing a basis for a common interpretation and uniform implementation of its provisions, and by ensuring that its provisions are duly taken into account in this Regulation and in the rules drawn up for its implementation;
e. to promote Community views regarding civil aviation safety standards and rules throughout the world by establishing appropriate cooperation with third countries and international organisations;
f. to provide a level playing field for all actors in the internal aviation market.

This list can be interpreted so to provide four main areas of activity for EASA; certification, research and safety analysis, rulemaking, and standardisation. With regard to cybersecurity, research and rulemaking have dominated the Agency's efforts so far under its competencies per Regulation 216/2008.

40 The +4 States are Iceland, Liechtenstein, Norway and Switzerland.

The scope of Regulation 216/2008 was consequently limited to safety and environment. As a result, this legal regime was only applicable to cybersecurity in as much as it related to these two areas.

While having a broad scope of competencies within its safety and environment mandate, there were other limitations imposed on this mandate.

First, the Regulation only applied to *civil* activities. Article 1(2)(a), echoing Article 3 of the Convention on International Civil Aviation of 1944 (Chicago Convention),[41] declared that the Regulation shall not apply to:

> products, parts, appliances, personnel and organisations referred to in paragraph 1(a) and (b) while carrying out military, customs, police, search and rescue, firefighting, coastguard or similar activities or services. The Member States shall undertake to ensure that such activities or services have due regard as far as practicable to the objectives of this Regulation.

With "aerodromes or part thereof, as well as equipment, personnel and organisations, referred to in paragraph 1(c) and (d), that are controlled and operated by the military" and "ATM/ANS, including systems and constituents, personnel and organisations, referred to in paragraph 1(e) and (f), that are provided or made available by the military" also being exempt per Article 1(2)(b) and (c) respectively. Therefore, anything related to cybersecurity within the scope of such activities or services fell outside the scope of Regulation 216/2008.

Second, Annex II of Regulation 216/2008 contains a list of aircraft types that were also outside the scope of the Agency's competencies and were instead regulated at a national level by the Member States. These included:

> a. historic aircraft [...];
> b. aircraft specifically designed or modified for research, experimental or scientific purposes, and likely to be produced in very limited numbers;
> c. aircraft of which at least 51% is built by an amateur, or a non-profit making association of amateurs, for their own purposes and without any commercial objective;
> d. aircraft that have been in the service of military forces, unless the aircraft is of a type for which a design standard has been adopted by the Agency;

41 Convention on International Civil Aviation (Chicago Convention), 61 Stat. 1180, 15 U.N.T.S. 295, 7 December 1944.

e. aeroplanes, helicopters and powered parachutes having no more than two seats, [...];
f. single and two-seater gyroplanes with a maximum take-off mass not exceeding 560 kg;
g. gliders with a maximum empty mass, of no more than 80 kg when single-seater or 100 kg when two-seater, including those which are foot launched;
h. replicas of aircraft meeting the criteria of (a) or (d) above, for which the structural design is similar to the original aircraft;
i. unmanned aircraft with an operating mass of no more than 150 kg;
j. any other aircraft which has a maximum empty mass, including fuel, of no more than 70 kg.

As a result, the scope of the Regulation was further limited. This meant that the competencies of the EU, including EASA, must be assessed below – within the context of these general limitations – in order to apprehend whether a cybersecurity measure could be taken under the aviation safety legal regime.

2.2.4 Regulation 216/2008 and Cybersecurity

Regulation 216/2008 did not mention the word 'cybersecurity' anywhere in its Articles or Annexes. However, despite the lack of clearly defined responsibilities for cybersecurity, this did not mean that Regulation 216/2008 was silent nor did it mean that the Agency did not have responsibilities in this area.

Annex Vb of Regulation 216/2008, which covers the Essential Requirements for ATM/ANS and Air Traffic Controllers, will serve as an example to highlight the Regulation's applicability to cybersecurity.[42]

> 2(c) Air traffic services
> (i) The data used as a source for the provision of air traffic services shall be correct, complete and current.

This provision was targeted towards the data provider, to ensure that the provided data was an accurate depiction of its subject. It is unlikely that cybersecurity was considered by the drafters when this provision was written, however, it has relevance for cybersecurity. This is because if the data was compromised due to a cyberattack, such as where the attacker deletes or modifies the data, this would result in it being neither correct nor complete. This shows that, if there was a link to safety and a relevant Article or Annex,

42 *See*, for addition information, the work carried out by Eurocontrol, Chapter III.6.3 and, for SESAR, Chapter II.6.3.

there would be no reason why Regulation 216/2008 could not be applicable to cybersecurity. Continuing within the scope of ATM/ANS, Regulation 2017/373 then covers 'data security', 'software integrity and security', and even uses the term 'cybersecurity' Annex III:

> ATM/ANS.OR.D.010 Security management
> [...]
> (d) Air navigation services and air traffic flow management providers and the Network Manager shall take the necessary measures to protect their systems, constituents in use and data and prevent compromising the network against information and cyber security threats which may have an unlawful interference with the provision of their service.[43]

This shows that elements of cybersecurity could be found in Regulation 216/2008 and it should, therefore, not be overlooked and dismissed as irrelevant. There are several other examples that could be taken which link cybersecurity to the safety Articles and Annexes of the Regulation. However, the above examination of Annex Vb 2(c)(i) is sufficient to show the relevance of Regulation 216/2008 to cybersecurity and negates the need for a full legal analysis of the text.

A Proposal was published by the Commission on 7 December 2015 which contained a draft text of a Regulation that would replace Regulation 216/2008. Article 76 of the draft text, titled 'Aviation Security', contained the term 'cybersecurity' which introduced the term into the EU aviation safety regime.[44] This would then give a clear mandate to the Agency with regard to cybersecurity which was lacking under Regulation 216/2008. This could be interpreted to mean that the Agency's cybersecurity activities conducted under Regulation 216/2008 were merely preparatory work in anticipation of the new Regulation. However, this is not the case, as demonstrated in the previous paragraph, as these activities were within the competencies of EASA as provided by Regulation 216/2008.

---

43 Commission Implementing Regulation (EU) No. 2017/373 of 1 March 2017 Laying Down Common Requirements for Providers of Air Traffic Management/Air Navigation Services and other Air Traffic Management Network Functions and their Oversight, Repealing Regulation (EC) No. 482/2008, Implementing Regulations (EU) No. 1034/2011, (EU) No. 1035/2011 and (EU) No. 2016/1377 and Amending Regulation (EU) No. 677/2011, OJ L 62, 8 March 2017, pp. 1-126.

44 Proposal for a Regulation of the European Parliament and of the Council on Common Rules in the Field of Civil Aviation and Establishing a European Union Aviation Safety Agency, and Repealing Regulation (EC) No. 216/2008 of the European Parliament and of the Council, COM/2015/0613 Final – 2015/0277 (COD), 7 December 2015.

A more in-depth analysis will, however, be undertaken in Chapter II.9 with regard to the replacement Regulation – Regulation 2018/1139 – as this is the new legal basis for aviation safety in the EU for the next decade.[45]

3 Early Steps Taken by the European Aviation Safety Agency

Two years after the adoption of the EU's cybersecurity strategy, the 2nd High-level Conference on the EU Cybersecurity Strategy took place on 28 May 2015 to provide "an overview of the state of play of the implementation of the five main priorities of the Strategy and showcase highlights of its main actions".[46] This event is seen as significant because it represents EASA's first external activity with regard to aviation cybersecurity, as depicted on the Agency's cyber-timeline.[47] From this date on, cybersecurity became more visible at an EU aviation level, with discussions and actions happening within an aviation context, as opposed to a general one. This does not mean that EASA was not active in this area prior to this date, however, it was a secondary consideration or termed differently. For example, in an Agency's 2009 report it referred to "security of using standardised data links for controlling [unmanned aircraft]"[48] and the Learjet Model 45 Certificate Data Sheet of 2014 utilised the term 'network security',[49] of which both terms could have cybersecurity connotations.

In mid-2016, EASA began to take steps towards directly regulating cybersecurity and to gaining a better understanding of cyber vulnerabilities in aviation. This step took place on 17 May 2016 when EASA launched Rulemaking Task RMT.0648 Aircraft Cybersecurity. The aim of this Rulemaking Task "is to mitigate the safety effects stemming from cybersecurity risks due to acts of unlawful interference with the aircraft on board electronic networks and systems".[50] This should eventually be realised by amending Certification

45 Regulation (EU) No. 2018/1139 of the European Parliament and of the Council of 4 July 2018 on Common Rules in the Field of Civil Aviation and Establishing a European Union Aviation Safety Agency, and Amending Regulations (EC) No. 2111/2005, (EC) No. 1008/2008, (EU) No. 996/2010, (EU) No. 376/2014 and Directives 2014/30/EU and 2014/53/EU of the European Parliament and of the Council, and Repealing Regulations (EC) No. 552/2004 and (EC) No. 216/2008 of the European Parliament and of the Council and Council Regulation (EEC) No. 3922/91, OJ L 212, 22 August 2018, pp. 1-122.

46 Commission, 'Second High-level Conference on the EU Cybersecurity Strategy', 28 May 2015, https://www.ec.europa.eu/digital-single-market/en/news/high-level-cybersecurity-conference-agenda.

47 EASA, 'Main EASA Activities', https://www.easa.europa.eu/easa-and-you/cyber-security/main-easa-activities.

48 EASA, Final Report of the Preliminary Impact Assessment on the Safety of Communications for Unmanned Aircraft Systems (UAS), Vol. 1, 8 December 2009, at 14.

49 EASA, Type-Certificate Data Sheet No. TCDS IM.A.020 for Learjet Model 45.

50 EASA, 'Rulemaking Programme 2016-2020', Final, 11 December 2015.

Specifications (CSs) and/or Acceptable Means of Compliance (AMC)[51] to, *inter alia*, CS-25 on large aeroplanes[52] and CS-29 on large rotorcraft.[53]

The next step took place on 26 May 2016 in Brussels, Belgium and was the First Meeting of Member States' Representatives of Cybersecurity in Aviation. During the meeting, the Agency presented its plans for cybersecurity as contained in its Roadmap and stressed its intention to work in close collaboration with Member States.[54]

Additionally, a workshop took place in Cologne, Germany between 13 and 14 July 2016. During this workshop, four mock cybersecurity crisis scenarios were enacted with the goal of identifying "the services and functions of a European Centre for Cybersecurity in Aviation (ECCSA) during the different events, its added value, possible constraints".[55] The conclusion of the workshop was axiomatic, in that there are both benefits and difficulties with sharing cyber related information, but there was also agreement on the need for a centralised coordination centre at European level through a ECCSA.[56]

These three early activities set the scene for EASA's cybersecurity actions as they cover regulation, cooperation with stakeholders and information sharing, which are at the centre of the EU's cybersecurity work. How these three have materialised will be shown in greater detail throughout Chapter II.

4 European Aviation Cyber Strategy – Bucharest Declaration

A high-level meeting took place between 8 and 9 November 2016 in Bucharest, Romania that brought together important aviation and cybersecurity stakeholders. This included representatives from the industry, Member States, European institutions including the Commission and EASA, academia and cybersecurity institutions. The aim of the meeting was to "discuss the aviation cybersecurity risk picture, actions already taken and ways forward"[57] due to the growing threat to aviation stakeholders from attackers and the

51 *See*, Chapter II.7.1.
52 Decision No. 2003/2/RM of the Executive Director of the Agency of 17 October 2003 on Certification Specifications, Including Airworthiness Codes and Acceptable Means of Compliance, for Large Aeroplanes («CS-25»), ED Decision No. 2003/2/RM, Final (2003) as Amended.
53 Decision No. 2003/16/RM of the Executive Director of the Agency of 14 November 2003 on Certification Specifications for Large Rotorcraft («CS-29»), ED Decision No. 2003/16/RM, Final (2003) as Amended. *See*, Chapter II.7.2.
54 *See*, Chapter II.7.2.
55 EASA, 'Workshop on the Roles and the Activities of a European Centre for Cybersecurity in Aviation, Cologne', 13-14 July 2016, https://www.easa.europa.eu/newsroom-and-events/news/workshop-roles-and-activities-european-centre-cybersecurity-aviation.
56 *See*, Chapter II.8.
57 Bucharest Declaration, Bucharest, Romania, Palace of the Parliament, 8-9 November 2016, at 1.

changing vulnerabilities. The result of the Conference, which contains the main conclusions of the two-day meeting, was the Bucharest Declaration.[58]

In appreciating the complexity of cybersecurity and cyberattacks, both generally and in aviation, the stakeholders concluded – in the Bucharest Declaration – that a layered defence approach is required to achieve a "comprehensive, seamless and adaptive level of protection of the European aviation system against cyber threats".[59] To achieve this, seven layers were identified as enabling objectives.

1. Coordination.
2. Sharing of information and reporting.
3. Regulations.
4. Risk assessments.
5. Cybersecurity promotion, awareness and preparedness.
6. Knowledge and foresight.
7. Commitment and resources.

Within these seven layers, the need for better regulation was identified and the role that EASA should play was depicted. As a result, this document provides an authoritative policy statement on EU aviation cybersecurity and must, therefore, be examined.

4.1 *Coordination*

The first layer specifies the need for greater coordination in three areas. First, there is a need for general cooperation across Europe in a coordinated way that encompasses the industry and institutional stakeholders, and also both aviation safety and security actors. There is acknowledgement that the EU aviation safety regime, through EASA, cannot be the only focus for cybersecurity, which is clearly reflected in Article 88 of Regulation 2018/1139.

Second, EASA is tasked with coordinating the European cooperation efforts so gaps and duplication are avoided, and to ensure that such efforts, where appropriate, collaborate with regional and international efforts.

Finally, there is an appreciation that cybersecurity is not just a regional concern and, because of this, NAAs must be involved. Therefore, NAAs are encouraged to engage with national level organisations responsible for cybersecurity in pursuance of greater cooperation.

58 *Ibid.*
59 *Ibid.*

It is unclear what the scope of this layer is as it does not specify what the cooperative activities should entail. Within the context of aviation cybersecurity, as highlighted below in Chapter II.4.2, cooperation is usually focused on information exchange. However, this is covered by the second layer so cooperation could be understood as something else.

The Bucharest Declaration focuses on cooperation between European aviation and cybersecurity stakeholders. However, it may be appropriate to extend the scope to entities outside of the European context, such as to the FAA, and to also include stakeholders whose scope is wider than just Europe, such as IATA and Airports Council International (ACI). The focus should be on the protection of all aviation stakeholders and not just on those confined to a certain geographical location.

To ensure the implementation of this envisaged cooperation, EASA was tasked with coordinating such cooperation. The appropriateness of EASA's leadership role is debatable because of the Agency's competing role as the EU aviation safety authority. This latter role endows the Agency with such responsibilities as compliance monitoring and enforcement. Thus, stakeholders may be reluctant to cooperate in areas that overlap with these responsibilities of EASA as they may face repercussions that affect them financially or operationally.

4.2 *Sharing of Information and Reporting*

There is often a lack of openness and transparency when it comes to perceived security threats, with States often treating or perceiving it – correctly in some cases – as a matter of national security. However, openness and transparency are more common in aviation safety. For example, the Confidential Safety Reporting system of EASA allows anyone to report information about a potential accident or incident in order to improve aviation safety.[60]

Information sharing is also found in aviation cybersecurity with the creation of the ECCSA.[61] The benefits of information sharing in aviation safety and cybersecurity are recognised, and this was adopted by the Bucharest Declaration as highlighted in the second layer: "The sharing of information about cyber threats, incidents and vulnerabilities is an essential element of the layered defence approach."[62]

60 EASA, 'Aviation Safety Reporting', https://www.easa.europa.eu/easa-and-you/safety-management/aviation-safety-reporting.

61 *See*, Chapter II.8.

62 Bucharest Declaration, 2016, at 2.

Information may be shared via two mechanisms. It can be done by way of a mandatory reporting system, such as with accident and incident reporting under Regulation 996/2010,[63] or via voluntary reporting based on trust, such as EASA's Confidential Safety Reporting system. It may be the case that not every entity wishes to cooperate so a legally mandated approach may be required to facilitate this activity. For example, exposing cyber vulnerabilities or a successful cyberattack can be extremely sensitive for companies, and they may not wish to share this information with other stakeholders, especially competitors. Therefore, mandating sharing may be the only way to achieve this objective. However, if the system is too invasive, this could result in a backlash from stakeholders, or in reportable cases being covered up or downplayed.

Once information has been shared to a central focal point, such as EASA, it should then be disseminated to relevant stakeholders that can benefit from the information. As a result, "Member States should initiate the development of cyber intelligence information and establish secure dissemination of threat-related information, towards those stakeholders who need to know".[64] Difficulties may arise in assessing *who* needs to know the information. For example, if EASA is the main focal point, relevant stakeholders may only include EU natural and legal persons. This would then leave out key manufactures such as Boeing (US), Bombardier (Canada) and Embraer (Brazil). Further, denying prospective participants access to the information sharing group could cause political and economic consequences. As this must be taken into consideration, entry requirements for entry into the cooperation were established.

Difficulties may also arise in assessing who needs to know *what* information. Requiring detailed information could make stakeholders hesitant or reluctant to share information, or too much information may result in 'information overload', both of which could result in important knowledge being overlooked. Alternatively, minimal sharing could have

63 Regulation (EU) No. 996/2010 of the European Parliament and of the Council of 20 October 2010 on the Investigation and Prevention of Accidents and Incidents in Civil Aviation and Repealing Directive No. 94/56/EC, OJ L 295, 12 November 2010, pp. 35-50.

64 Bucharest Declaration, 2016, at 2.

safety implications as not every bit of relevant information may be circulated. There are clear practical issues with information sharing that need to be considered.

There are also legal issues that need to be considered. For example, the EU, including EASA, is governed by Regulation 2018/1725 on data protection.[65] For other stakeholders, the EU rules on data protection found in Regulation 2016/679, or more commonly the GDPR, on the processing and free movement of personal data will also be applicable.[66] As a result, there are EU-wide rules on the hosting and distribution of information, which will control how cybersecurity information is treated.

Regulation 2018/1139 also imposes obligations on the Agency with regard to:

1. Article 73: Protection of the Source of Information.
2. Article 74: Repository of Information.
3. Article 123: Security Rules on the Protection of Classified and Sensitive Non-Classified Information.
4. Article 132: Processing of Personal Data.

The Agency must, consequently, consult its own establishing Regulation, as well as other sources of EU law.

An additional legal consideration for the EU is found in Regulation 1049/2001 on access to documents.[67] Even if the information sharing system does not legally mandate information exchange, the EU may still be obliged to provide documents under this Regulation. Therefore, there is a wider legal context that must be appreciated when hosting and sharing data.

65 Regulation (EU) No. 2018/1725 of the European Parliament and of the Council of 23 October 2018 on the Protection of Natural Persons with Regard to the Processing of Personal Data by the Union Institutions, Bodies, Offices and Agencies and on the Free Movement of Such Data, and Repealing Regulation (EC) No. 45/2001 and Decision No. 1247/2002/EC, OJ L 295, 21 November 2018, pp. 39-98. *See*, for data protections relation to EASA, Regulation 2018/1139, Arts 74(5) and 132(2), and Recital 50. *See also*, for the rules applicable prior to the entry into force of Regulation 2018/1725, of which some of the EU's cybersecurity activities, including EASA's, Regulation (EC) No. 45/2001 governed these activities. *See*, Regulation (EC) No. 45/2001 of the European Parliament and of the Council of 18 December 2000 on the Protection of Individuals with Regard to the Processing of Personal Data by the Community Institutions and Bodies and on the Free Movement of Such Data, OJ L 8, 12 January 2001, pp. 1-22.

66 Regulation (EU) No. 2016/679. *See*, Regulation 2018/1139, Arts 74(5) and 132(1), and Recitals 31, 33 and 49.

67 Regulation (EC) No. 1049/2001 of the European Parliament and of the Council of 30 May 2001 Regarding Public Access to European Parliament, Council and Commission Documents, OJ L 145, 31 May 2001, pp. 43-48.

Despite these issues, it is clear that information sharing is going to be an important element of aviation cybersecurity and that information needs to have a two-way flow: (1) From the information owner to a central location and then (2) from the central location to other stakeholders for use.[68]

4.3 Regulations

It was declared in the third layer that regulations should be developed in the area of aviation cybersecurity. Four guiding principles should steer risk and performance-based sectorial regulations, so as to provide a level playing field and a balanced sharing of risk management. The regulations should:

1. stay relevant with the evolution of threats and technologies,
2. be consistent and complementary with international efforts,
3. be consistent and complementary with EU law,
4. take into consideration existing national rules, and
5. utilise industry standards where applicable.[69]

It is essential that the law is pragmatic and allows for changes to be made when required, as threats and vulnerabilities emerge, evolve and disappear. Therefore, a static regulatory system would not be appropriate as this would make the law liable to becoming inefficient or redundant. However, not every aspect will require the same level as pragmatism as, for example, the list of victims and their targets assets (i.e. aircraft, engines) will not evolve as quickly as the available attack methods. As result, the regulation should take this into consideration.

The EU aviation legal system has emerged within the context of international air law, such as that developed under the auspices of ICAO and national law. This layer clearly acknowledges this context, accounting for the limitations of EU law and the necessity of harmonisation. Harmonisation is also promoted between the EU aviation rules and the wider EU legal context, and also at a national level. This is necessary to appreciate as aviation specific rules are not capable of encapsulating all of the necessary rules to regulate cybersecurity. Therefore, the specialised rules can tackle specialised topics, as being developed under the EU's Cybersecurity Strategy and Digital Single Market.

---

68 *See*, for general information, E. Hoeven, 'Cybersecurity: Information-Sharing is Critical to Building Aviation System Resiliency', 71(4) *ICAO Journal* 38-40 (2016).

69 Bucharest Declaration, 2016, at 2.

Finally, the inclusion of industry standards can be seen in other areas of aviation regulation, such as with unmanned aircraft.[70] The use of industry standards in European aviation regulation is increasing as it allows EASA to utilise industry expertise. This is common with cybersecurity as demonstrated in Chapter II.7.2.1 on certification.

The proclamation of these five guiding principles and their actual implementation are a different matter, so only an analysis of new regulations will give credibility to the alignment of theory and practice.

4.4 *Risk Assessments*

There is an acknowledgement that risk assessments should be coherent, comparable, share common risk criteria, and share terminologies and methodologies. No information is given in the document to help the reader understand what the risk assessments should check for, what should be done with the results of the risk assessment, how the stated requirements should be achieved or the legal basis for such actions. Regarding content, there are several documents that can assist, such as:

1. ICAO Doc 9985 on Air Traffic Management Security Manual.
2. EUROCAE ED 202A and 203A.[71]
3. Part 21, which mandates security risk assessments for products and parts.

Regarding the legality, risk assessments can be found in several sources of law, such as the NIS Directive and Regulation 300/2008 on aviation security,[72] so clarification is required to determine whether this layer refers to one of these sources of law, both or whether it is a general statement that calls for harmonisation when and wherever a risk assessment should arise in aviation cybersecurity.

4.5 *Cybersecurity Promotion, Awareness and Preparedness*

Although the creation of new regulations is an important step in enhancing cybersecurity in aviation, which has been duly noted in Chapter II.4.3, information about this must be distributed and accessible to those concerned. Therefore, the fifth layer notes that cybersecurity is a mind-set that should be shared by all stakeholders. Stakeholders should

70 *See*, Opinion No. 01/2018, Introduction of a Regulatory Framework for the Operation of Unmanned Aircraft Systems in the 'Open' and 'Specific' Categories, 6 February 2018.

71 These Standards will be used to form some of the AMC listed for RMT.0648.

72 Regulation (EC) No. 300/2008 of the European Parliament and of the Council of 11 March 2008 on Common Rules in the Field of Civil Aviation Security and Repealing Regulation (EC) No. 2320/2002, OJ L 97, 9 April 2008, pp. 72-84.

promote cybersecurity and train necessary persons to develop awareness and preparedness, and this training should be consistent with cross domain exercises that are being organised. This is an important complement to regulatory efforts, but it also supports cybersecurity more generally as it covers training which helps support different control types[73] and control modes,[74] and this is vital for effective cybersecurity.

4.6 Knowledge and Foresight

There is a consensus among stakeholders that there is a lack of knowledge in this area, promulgated by the novelty of aviation cybersecurity. Therefore, the advancement of knowledge is promoted by the sixth layer which supports research and studies that "expand the collective knowledge on current and future cyber risks, and on the ability to protect civil aviation systems against cyber threats".[75] This is a mature approach to regulating and should be commended as it will enhance the quality of the rules that are produced. However, as cyber is a growing and evolving threat, there must be a balancing act between conducting research, which provides an understanding of the topic but can also be time consuming and expensive, and the implementation of rules that should protect aviation stakeholders.

4.7 Commitment and Resources

The final layer declares that all stakeholders should allocate resources, whether human, monetary or technical, as appropriate for aviation cybersecurity. The inclusive nature of this allows for all aspects to be covered under the heading 'resources' so it should not leave any gaps. However, the subjective nature of this layer, provided through the wording 'appropriate level', could result in a disproportionate allocation of resources. The resources held by one stakeholder differs from that of another, each may have different priorities and they may have different expertise. Therefore, one stakeholder may set the subjective level of 'appropriateness' much higher or lower than another. This may then have consequences for cybersecurity if the level of what is deemed appropriate is too low.

Only those stakeholders with the sufficient amount of resources will be able to implement appropriate cybersecurity measures. This could then lead to some stakeholders being more vulnerable than others, as they would be unable to dedicate the necessary resources

73 *See*, Conclusion 3.2.
74 *See*, Conclusion 3.3.
75 Bucharest Declaration, 2016, at 3.

to a certain vulnerability. This could funnel cyberattacks towards less resource intensive stakeholders as attackers may choose to target the weaker links in the aviation chain.

It may be appropriate for regulations to acknowledge the differences between stakeholders and prescribe measures that mitigate any negative results stemming from varied resource allocation as this would help reduce the vulnerabilities of less capable stakeholders.

The establishment of an aviation cybersecurity fund to support stakeholders that struggle to meet the required levels of cybersecurity could be a solution. This is a feasible as the Commission has discussed funding within general discussions of cybersecurity.[76]

As no stakeholder has unlimited resources, the guiding principle needs to be finding a balance between resource expenditure and the cost incurred resulting from a successful exploitation by the attacker.

4.8 *Next Steps*

The Declaration ends by stating that a European Strategic Coordination Platform (ESCP) shall be set up to coordinate the definition and implementation of a European strategy for Cybersecurity in Aviation.[77] The first meeting was held in May 2017 at EASA's headquarters in Cologne, Germany.

4.9 *Bucharest Conference in Context*

The Bucharest Conference draws parallels with the Riga Conference which took place during the Latvian Presidency of the Council of the EU in March 2015. The Riga Conference was organised by the Ministry of Transport of Latvia and the Civil Aviation Agency of Latvia, in cooperation with the European Commission in order to bring together key European stakeholders involved with unmanned aircraft. The result of the conference was the signing of the Riga Declaration[78] which was endorsed by the Transport Commissioner, and the Directorate-General for Mobility and Transport of the Eu-

76 Commission, 'EU Budget: Commission Proposes €9.2 Billion Investment in First Ever Digital Programme', Press Release, 6 June 2018.

77 *See*, Chapter II.6.5.

78 Riga Declaration on Remotely Piloted Aircraft (Drones), 'Framing the future of aviation', Riga, Latvia, 6 March 2015. *See*, V. Correia and N. Rouissi, 'The European Union and Civil Drones: The Riga Declaration and the Future of the European RPAS Industry', *in* B. Scott (ed.), *The Law of Unmanned Aircraft Systems: An Introduction to the Current and Future Regulation of UAS Under National, Regional and International Law*, (Wolters Kluwer, 2016).

ropean Commission. The Declaration "set out the general principles that will serve as a guide to the regulatory process in Europe and called for the Agency to develop the rules".[79] This method of working is becoming more prevalent in the EU for emerging aviation areas, however, there is a noticeable difference between the Bucharest Declaration and the Riga Declaration.

The Riga Declaration gave a strong political statement in favour of the direction regulation should go and provided endorsement from the Stakeholders for the EU to undertake the regulation of unmanned aircraft. This is not the case for cybersecurity as regulatory steps had already taken place prior to the Bucharest Conference as outlined in the opening page of the Bucharest Declaration.

1. The Commission and EASA, in cooperation with Member States and Industry, had developed a Roadmap.[80]
2. The inclusion of Article 88 in the revision of 216/2008 proposal.[81]
3. The Agency was in the process of implementing an ECCSA.[82]

Therefore, the importance of the Bucharest Declaration as a policy tool, and even as a guide for the regulation of aviation cybersecurity is arguably less influential than the Riga Declaration within the unmanned aircraft domain. Nevertheless, it should not be dismissed, as it provides wide endorsement for the work, including planning, that has already been carried out by the Commission and EASA, as well as a good indication of what still needs to be done.

4.10 Transport Cybersecurity Conference

On 23 January 2019, ENISA organised the First Transport Cybersecurity Conference which took place in Lisbon, Portugal. This was done in cooperation with the Commission (i.e. DG-MOVE), EASA, the EU Agency for Railways and the European Maritime Safety Agency. The aim of the conference was to discuss cybersecurity in a coordinated manner between the different transport sectors.[83] EASA's Flight Standards Director, Jesper Rasmussen, stated:

79 *See*, G. Boccardo, 'European Aviation Safety Agency', *in* B. Scott (ed.), 2016.

80 *See*, Chapter II.7.2.2.

81 *See*, Chapter II.9.5. Proposal, COM/2015/0613 Final – 2015/0277 (COD), 7 December 2015.

82 *See*, Chapter II.8.

83 ENISA, '1st Transport Cyber Security Conference', https://www.enisa.europa.eu/events/first-transport-cyber-security-conference. *See*, for the results of an earlier meeting of European stakeholders, ENISA, Cyber Europe 2018: After Action Report: Findings from a Cyber Crisis Exercise in Europe, December 2018.

> To deter potential cyberattacks targeted at the transport system, cooperation is vital among all partners. Today's discussions have highlighted that aviation is probably in front among the transport modes to launch concrete initiatives. EASA is working on ensuring a high and equal level of protection in the entire EU aviation system through a coming regulation of cyber security. And EASA is together with industry setting up the institutional framework to facilitate exchange of sensitive information on cyber security so all can learn and improve.[84]

The conclusions of the conference can be summarised into four key topics:
1. The wider regulatory context must be considered: EASA is not the only regulator, so consideration for other regulatory systems, such as the EU's rules on network and information systems (e.g. NIS Directive), is needed to avoid duplication and contradictions.
2. All relevant stakeholders should be involved: All relevant stakeholders should be involved, without sectoral or geographical limitations.
3. Information sharing: There is a need for continuous and growing levels of information sharing.
4. Training: People must be trained in cybersecurity in order to fill demand.

These four points, while already featuring prominently in the Bucharest Declaration, are important and vital to reiterate.

5 Information Sharing Activities Post Bucharest

The Agency is aware that no single entity has all of the necessary information to implement effective cybersecurity measures and mitigate cyber vulnerabilities. As a result, subsequent to the Bucharest Declaration, EASA has been active with propagating information acquisition and dissemination.

On 12 December 2016, the Agency launched the Impact Assessment of Cybersecurity Threats project.

> The project aims at developing a comprehensive knowledge base for evaluating the safety impact of a number of potential threats to critical aircraft systems,

84 EASA, '1st Transport Cybersecurity Conference held in Lisbon', https://www.easa.europa.eu/newsroom-and-events/news/1st-transport-cybersecurity-conference-held-lisbon.

> consolidating the methods and techniques used, as well as to enable the EU harmonisation of the associated risk management process.[85]

It is an external project in which the Agency has opened a public tender under Directive 2014/24/EU.[86] It is then planned that:

> the project encompasses the preliminary risk assessment at system and aircraft levels for potential cyberattacks to the following commercial transport aircraft (CAT) aircraft systems: flight management system (FMS), global navigation satellite system (GNSS) receiver, including [Ground Based Augmentation System] (GBAS) and satellite-based augmentation system.[87]

The output of the project is expected to consist of "a series of comprehensive guidelines for aircraft/system development to support the preliminary security risk assessment process along EUROCAE ED-202A airworthiness security process specification standard".[88] The results have not been made public.

EASA and the European Organisation for Civil Aviation Equipment (EUROCAE) also organised a Cybersecurity in Aviation workshop in Brussels, Belgium on 31 May 2017 to "discuss current activities and future regulatory and standardisation needs with key stakeholders from industry, airspace users, Member States, European Institutions, academia".[89] The content of the workshop consisted of the core content of Chapter II of this study so it does not require further elaboration. Nevertheless, it should be noted that the workshop demonstrates EASA's willingness to keep key stakeholders up-to-date with its activities and also to receive feedback from relevant sources. This provides clear adherence to the Bucharest Declaration and confirms the Agency's mature approach to rule-making.

Further, between 5 and 6 June 2018, EASA organised the 1st OPTICS2 Workshop – Aviation Cybersecurity which brought "together a range of experts from industry and

---

85 Commission, 'Germany-Cologne: Impact Assessment of Cybersecurity Threats', Contract Notice: Services, 2016/S 144-259890.

86 Directive No. 2014/24/EU of the European Parliament and of the Council of 26 February 2014 on Public Procurement and Repealing Directive No. 2004/18/EC, OJ L 94, 28 March 2014, pp. 65-242.

87 Commission, 2016/S 144-259890.

88 *Ibid.*

89 EASA, 'Main EASA Activities', https://www.easa.europa.eu/easa-and-you/cyber-security/main-easa-activities.

regulation to consider both existing and potential future cyber risks and identify a top ten list of both urgent and mid-term research that is required to maintain a safe and secure air transport system, today and in the future".[90] This further demonstrates EASA's desire to cooperate with the industry to share ideas, skills and information.

6 EUROPEAN AVIATION CYBER STRATEGY – KRAKOW DECLARATION

6.1 *High Level Conference on Cybersecurity*

The Polish Ministry of Infrastructure and Construction, together with EASA and the Polish CAA, organised an international High Level Conference on Cybersecurity in Civil Aviation to discuss the progress of the measures already taken "and to establish a ground for the future European strategy for Cybersecurity in Aviation and the Cybersecurity Road Map that will define the future EU actions that have to be undertaken in order to ensure a secure environment for aviation".[91]

The Krakow Conference built upon the discussions held during the Bucharest Conference, as well as the High Level Conference on Drones held in Warsaw, Poland on 23 to 24 November 2016.[92] As a result, there is a historical context to the Krakow Conference which is reflected in the resulting Krakow Declaration which contains 22 points.[93] The key themes of the 22 points are discussed below so as to provide an insight into the document's relevance for the regulation of aviation cybersecurity in the EU.

6.2 *Role of the European Aviation Safety Agency*

At the rudimentary rulemaking level, the Declaration calls upon the Council and the European Parliament to conclude the replacement for Regulation 216/2008 and for it to clearly define the role of EASA with regard to cybersecurity. This has been achieved and the subject of this action will be discussed in Chapter II.8.

---

90 EASA, '1st OPTICS2 Workshop – Aviation Cybersecurity', https://www.easa.europa.eu/newsroom-and-events/events/1st-optics2-workshop-aviation-cybersecurity; and OPTICS2, '1st OPTICS2 Workshop on Aviation cybersecurity (5-6 June 2018, EASA Cologne, Germany)', http://www.optics-project.eu/workshop1/.

91 Krakow Declaration, High Level Conference Cybersecurity in Civil Aviation Krakow Declaration, Krakow, Poland, 8-9 November 2017, at 1.

92 Warsaw Declaration, 'Drones as a Leverage for Jobs and New Business Opportunities', Warsaw, Poland, 24 November 2016.

93 Krakow Declaration, 2017.

Additionally, other regulations are being discussed as the Declaration calls upon the Commission and EASA to develop and adopt Implementing Regulations addressing Cybersecurity. This has yet to happen, however, steps have been taken by EASA under RMT.0720 Cybersecurity Risk.[94]

The Declaration also focused on the activities of EASA away from its direct rulemaking responsibilities. This includes, *inter alia*, its work on ECCSA,[95] its cooperation with the Computer Emergency Response Team – EU (CERT-EU)[96] and its support to other Regional Safety Oversight Organisations (RSOOs).[97] No evidence of support given to RSOOs on cybersecurity matters could be found and RSOOs are not currently active in this area, so EASA may have to act as a catalyst or wait for the RSOOs to take up cybersecurity matters.

The Declaration encourages EASA to ensure that the European Plan for Aviation Safety (EPAS) appropriately reflects all the necessary actions required at European level.[98]

Each area of rulemaking covered by the Declaration, as outlined above, has begun to be implemented. However, implementation may pre-date the Declaration, such as RMT.0720, so the document can be considered as a reminder to stakeholders of the activities that still need to be completed, rather than as an advertisement for future activities.

6.3 *Role of Other Stakeholders*

EASA and the aviation regulatory system were not the only areas to receive attention from the Declaration.

First, the role of the European Union Agency for Network and Information Security (ENISA)[99] was acknowledged as important for "aligning the responses of cyber threats against essential services, including civil aviation, at European level".[100]

94 *See*, Chapter II.7.2.4.
95 *See*, Chapter II.8.
96 *See*, Chapter II.8.2.
97 *See*, M. Ratajczyk, *Regional Aviation Safety Organizations: Enhancing Air Transport Safety Through Regional Cooperation*, (Wolters Kluwer, 2015); and P. Mendes de Leon, 2017, Ch. 7; and Chapter III.7.
98 *See*, Chapter II.7.2.
99 *See*, Chapter II.7.2.1.
100 Krakow Declaration, 2017, at 1.

ENISA is an Agency of the EU tasked to improve network and information security in the EU. It was established in 2004 by Regulation 460/2004,[101] which is no longer in force, with a periodically renewed mandate. The current ENISA mandate, which will expire in June 2020, is set out in Regulation 526/2013.[102] The objectives of ENISA are found in Article 2 of Regulation 526/2013.

1. The Agency shall develop and maintain a high level of expertise.
2. The Agency shall assist the Union institutions, bodies, offices and agencies in developing policies in network and information security.
3. The Agency shall assist the Union institutions, bodies, offices and agencies and the Member States in implementing the policies necessary to meet the legal and regulatory requirements of network and information security under existing and future legal acts of the Union, thus contributing to the proper functioning of the internal market.
4. The Agency shall assist the Union and the Member States in enhancing and strengthening their capability and preparedness to prevent, detect and respond to network and information security problems and incidents.
5. The Agency shall use its expertise to stimulate broad cooperation between actors from the public and private sectors.

ENISA has broad competencies to assist the EU, including EASA, Member States, the private sector and the public sector with cybersecurity. There is a proposal by the EU to repeal and replace Regulation 526/2013 with a Regulation dubbed the 'Cybersecurity Act'.[103] The proposal gained political agreement on 10 December 2018 by the European Parliament, the Council of the European Union, and the European Commission. The regulation, which is referred to as the 'Cybersecurity Act' will have the following outcomes:

---

101 Regulation (EC) No. 460/2004 of the European Parliament and of the Council of 10 March 2004 Establishing the European Network and Information Security Agency, OJ L 77, 13 March 2004, pp. 1-11.

102 Regulation (EU) No. 526/2013 of the European Parliament and of the Council of 21 May 2013 Concerning the European Union Agency for Network and Information Security (ENISA) and Repealing Regulation (EC) No. 460/2004, OJ L 165, 18 June 2013, pp. 41-58.

103 Proposal for a Regulation of the European Parliament and of the Council on ENISA, the 'EU Cybersecurity Agency', and Repealing Regulation (EU) 526/2013, and on Information and Communication Technology Cybersecurity Certification ('Cybersecurity Act'), Brussels, Belgium, COM (2017) 477 Final, 2017/0225 (COD), 13 September 2017.

1. ENISA will receive a permanent mandate with more human and financial resources;
2. ENISA will increase its support to EU Member States, in order to improve capabilities and expertise, notably in the areas of cyber crisis coordination and the prevention of and response to cyber-incidents;
3. Within the Cybersecurity Certification Framework, ENISA will have market-related tasks, notably by preparing European cybersecurity certification schemes with the expert assistance and close cooperation of national certification authorities and industry;
4. ENISA will strengthen its support to Member States and the EU institutions in the development, implementation and review of general cybersecurity policy.[104]

The Cybersecurity Act was published in the EU Official Journal on 7 June 2019 and is called Regulation 2019/881 on ENISA (the European Union Agency for Cybersecurity) and on information and communications technology cybersecurity certification. It repeals Regulation 526/2013.

ENISA was set up in 2004 to contribute to the overall goal of ensuring a high level of network and information security within the EU. ENISA helps the Commission, the Member States and the business community to address, respond and to prevent NIS problems. The main activities run by ENISA include:

1. Collecting and analysing data on security incidents within Europe and emerging threats.
2. Promoting risk assessment and management methods in order to enhance capability to deal with information security threats.
3. Running European cyber exercises.
4. Supporting CERTs cooperation in the Member States.
5. Raising awareness and levels of cooperation between different stakeholders involved in information security.

In order to boost the overall level of online security in Europe, ENISA organises each October the European Cybersecurity Month awareness campaign, with the support of NIS contact points in all Member States.

104 ENISA, EU leaders agree on ground-breaking regulation for cybersecurity agency ENISA, https://www.enisa.europa.eu/news/enisa-news/eu-leaders-agree-on-ground-breaking-regulation-for-cybersecurity-agency-enisa.

Second, the relevance of EU law was raised as the Declaration calls on EU Member States to take national measures in line with the NIS Directive. Thus, there is a promotion of harmonisation between regional EU law and the national law of Member States. This will have relevance for aviation stakeholders as, for example, aviation is a subject of the NIS Directive which must be implemented at a national level, which has so far had a limited level of success with low implementation. For example, the Commission sent "a letter of formal notice to 17 Member States to fully transpose into national laws the first piece of EU-wide legislation on cybersecurity".[105]

Third, the Declaration recognises the valuable work of ECAC in the area of cybersecurity, such as the updated ECAC Document 30, Chapter 14 on Cybersecurity and its supporting Guidance Material. This is again relevant for aviation cybersecurity regulation and will, therefore, be discussed in Chapter II.10.2.

Fourth, other stakeholders were noted, including airports, ground handling operators, maintenance organisations and air navigation service providers, and their role in developing information security management systems in accordance with specific procedures and appropriate standards. Therefore, the Declaration acknowledges the multifaceted characteristic of the aviation stakeholder landscape and its relevance for cybersecurity. This point is furthered as it was "recognised that cybersecurity is an interdisciplinary problem in transport that has its challenges in aviation, but also in shipping, rail and road transport".[106] Aviation should not be the sole focus of the EU efforts, as other transport methods and sectors are also vulnerable to cyberattacks, and aviation cybersecurity efforts should not be conducted in isolation as valuable lessons may be taken from the efforts of transport methods and sectors. This is also important for multinational transport, such as with cargo, which may utilise multi-modal transportation. Therefore, wider cooperation is required.

Fifth, the Declaration invites all relevant research actors, notably the Single European Sky ATM Research Joint Undertaking (SESAR),[107] Shift2Rail, national research institutions and industry to combine efforts. SESAR Joint Undertaking is a public-private partnership (PPP) that manages the development phase of SESAR. This will provide Europe with a high-performance ATM infrastructure and is involved with cybersecurity.

105 The 17 Member States: Austria, Bulgaria, Belgium, Croatia, Denmark, France, Greece, Hungary, Ireland, Latvia, Lithuania, Luxembourg, the Netherlands, Poland, Portugal, Romania, and Spain. Commission, July Infringements Package: Key Decisions, Fact Sheet, 19 July 2018.

106 Krakow Declaration, 2017, at 2.

107 *See*, SESAR, 'Home', http://www.sesarju.eu/.

> The SESAR Joint Undertaking (SJU) study on a cybersecurity strategy in aviation concluded on a number of recommendations which need to be followed in the development and deployment of the future air traffic management system. Currently, SJU is preparing internal standards to ensure that the risks related to cybersecurity are appropriately addressed in all projects. Cybersecurity is now an integral part of the new EU ATM Master Plan and of the SESAR 2020 Work programme. In addition, the SESAR Deployment Manager addresses cybersecurity in SESAR implementation activities following the Deployment Programme specific requirements.[108]

Therefore, the modernisation of ATM infrastructure in the EU will mean that cybersecurity is taken into account from design to deployment.[109]

This is in line with the strong cooperative and information sharing push found in the Bucharest Declaration.

6.4 *Non-Regulatory Solutions*

In addition to regulatory steps, the Declaration also sees the need for non-regulatory action. First, this is expressed as it calls for "a stronger partnership between regulators, operators, service providers, and [the] manufacturing industry, in particular within the ESCP, where EASA welcomes and supports the Industry to come up with standards".[110]

Second, the Declaration calls for continuing research and development activities for cybersecurity as part of the overall cybersecurity strategy of the EU.

Third, the Research Framework Programme – FP9,[111] which will replace Horizon 2020,[112] should make aviation-relevant subjects a high priority.

108 Commission, Commission Staff Working Document Impact Assessment, Accompanying the Document, Proposal for a Regulation of the European Parliament and of the Council on ENISA, The 'EU Cybersecurity Agency', and Repealing Regulation (EU) 526/2013, and on Information and Communication Technology Cybersecurity Certification ('Cybersecurity Act'), SWD (2017), 500 Final, Part 6/6, Brussels, Belgium, 13 September 2017, at 110.

109 *See, e.g.*, SESAR, 'New Study Reports on the Cyber-Security of SESAR's Airport Operations Centre', https://www.sesarju.eu/newsroom/all-news/new-study-reports-cyber-security-sesar%E2%80%99s-airport-operations-centre.

110 Krakow Declaration, 2017, at 2.

111 Horizon 2020, which is the the 8th EU Framework Programme for Research and Innovation, runs until the end of 2020. The preparations for the future framework programme, which is called 'FP9', are currently ongoing.

112 Commission, 'Horizon 2020', https://www.ec.europa.eu/programmes/horizon2020/.

Horizon 2020: It is the largest ever EU Research and Innovation programme with circa EUR 80 billion of funding available between 2014 and 2020. It has taken numerous theoretical ideas, including within the aviation sector, and delivered them to market.

If this is adopted, then significant developments can be expected.

These three areas repeat earlier calls for increased cooperation and research, as well as noting the relevance of industry standards to complement more stringent regulations. Thus, non-legal methods are very important in the EU's aviation cybersecurity strategy and will play an important role in managing cybersecurity.

6.5 European Strategic Coordination Platform

The ESCP, which is mentioned in four separate points of the Declaration, was set up by the industry to do a similar task to the ECCSA.[113] As a result, questions about the duplication of work and harmonisation can be raised. However, it should be noted that ESCP is concerned with strategic operations and the ECCSA is focused on operational activities, so each has a different focus.

The group consists of key industry stakeholders, States and EU institutions, which substituted an existing informal group which involved, *inter alia*, manufacturers and airlines. While information about the ESCP is limited, it did convene for the first time in 2017 and provided an opportunity for the EU to have a dialogue with experts with operational knowledge to assist with the drafting of EU law.

The Declaration's four separate points on the ESCP note its importance in advancing aviation cybersecurity, call for further cooperation between ESCP and all relevant stakeholders, stress the need to conclude its engagement phase before 2018 ends, and acknowledge the need to assess the progress made in aviation cybersecurity with the help of ESCP within 12 months.

113 J. Anton, EASA, 'Cybersecurity; An EASA Perspective on Developments and Challenges', Presentation: IATA Paperless Conference 2017, 13-14 November 2017. *See*, Air Traffic Management Net, 'European Fragmentation Hampering Cyber Effort', 28 March 2017, http://www.airtrafficmanagement.net/2017/03/european-fragmentation-hampering-cyber-effort/.

7 European Aviation Safety Agency Rulemaking Programme

7.1 Rulemaking Competencies of European Aviation Safety Agency

The EU, acting through EASA, has been tasked by the Bucharest Declaration and Krakow Declaration to develop regulations, facilitate cooperation and information sharing, carry-out research, and promote cybersecurity. In pursuance of this, the Agency has implemented several rulemaking efforts since 2016, which must be analysed to gauge how cybersecurity is being regulated at an EU level.

There is an established process that is set out in the Agency's foundational Regulation[114] and internal documents[115] that facilitate the rulemaking duties of EASA. The first step in the rulemaking process requires the Executive Director (ED), currently Patrick Ky, to establish an annual 5-year Rulemaking Programme which outlines the tasks to be completed in the opening year, while also laying out the rulemaking plans for the following four years. The content of the programme results from consultation with the Member States, stakeholders and the Commission, and is based on the Agency's competencies, as well as necessity.

The Rulemaking Programme contains six different categories of tasks that can be adopted by the Agency.

1. Safety Promotion (SPT).
2. Research Actions (RES).
3. Rulemaking Task (RMT).
4. Focused Attention Topic (FOT).
5. Member States Tasks (MST).
6. Evaluation Task (EVT).

The first three are relevant for cybersecurity and will be the focus of this subchapter.

> Safety Promotion (SPT): This incorporates "a set of means, processes and procedures that are used to develop, sustain and improve aviation safety through awareness raising and changing behaviours".[116]

---

114 *See*, for legal basis of cited Rulemaking Programmes, Regulation 2018/1139, Art. 76.

115 EASA, Management Board Decision 18-2015, 'Management Board Decision 18-2015 of 15 December 2015 Replacing Decision 01/2012 Concerning the Procedure to be Applied by the Agency for the Issuing of Opinions, Certification Specifications, Acceptable Means of Compliance and Guidance Material ('Rulemaking Procedure')', 15 December 2015.

116 *See*, EASA, 'Safety Promotion', https://www.easa.europa.eu/easa-and-you/safety-management/safety-promotion.

Research Actions (RES): EASA has the competency to launch and finance research projects within its safety and environment mandate.
Rulemaking (RMT): The Agency has the competency to develop different types of regulatory material.

As a part of its rulemaking tasks, the Agency is competent to issue Opinions to the Commission, which are then assessed and adopted, upon approval, as new or amended EU secondary legislation, notably Commission Regulations.[117] This form of rulemaking by the Agency is unofficially referred to as 'hard law'. In addition, the Agency shall adopt, what is unofficially referred to as 'soft law', three types of non-binding materials.

Certification Specifications (CSs): These are used to establish the certification basis for applications.
Acceptable Means of Compliance (AMC): This is a standard that states how compliance with Regulation 216/2008 and its Implementing Regulations is to be achieved.
Guidance Material (GM): This is material that assists with the understanding of a requirement or specification, and supports interpretation of Regulation 216/2008, its Implementing Regulations and AMCs.

EASA can also cooperate with the setting of international standards and, when requested, assists third countries[118] with their own internal rulemaking.

Management Board Decision (MBD) 18-2015 then contains the procedure that the Agency should follow when developing the above-mentioned rules, including Terms of Reference (ToR), Notices of Proposed Amendment (NPA) and Advanced Notices of Proposed Amendment (A-NPA).[119]

117 *See*, P. Mendes de Leon, 2017, at 329-330.

118 The term 'third countries', as used by EASA in this context, will be used here as opposed to 'third States' for consistency reasons. Commission Regulation (EU) No. 452/2014 of 29 April 2014 Laying Down Technical Requirements and Administrative Procedures Related to Air Operations of Third Country Operators Pursuant to Regulation (EC) No. 216/2008 of the European Parliament and of the Council Text, OJ L 133, 6 May 2014, pp. 12-26. It has been amended by Commission Regulation (EU) No. 2016/1158 of 15 July 2016 Amending Regulation (EU) No. 452/2014 as Regards the Deletion of Templates for the Authorisations Issued to Third Country Operators and for the Associated Specifications OJ L 192, 16 July 2016, pp. 21-22.

119 EASA, Management Board Decision 18-2015, 2015.

7.2 Rulemaking Programme and Cybersecurity

Cybersecurity is first referenced in the 2016-2020 Rulemaking Programme which was officially adopted by the ED Decision 2015/236/ED on 11 December 2015 and supersedes previously published programmes. The 2016-2020 Rulemaking Programme contained two cybersecurity activities.

1. RMT.0648 – Aircraft Cybersecurity.
2. SPT.071 – Cybersecurity Road Map.

Rulemaking Programme for 2017-2021 then builds upon the earlier programme and stresses, in light of the growing threat that cybersecurity presents to aviation, the need for up-to-date information and the necessity for stakeholders to share such information: "Updated information and the sharing of knowledge is of paramount importance".[120] The programme contains three activities specific to cybersecurity.

1. RMT.0648 – Aircraft Cybersecurity.
2. SPT.071 – Cybersecurity Road Map.
3. SPT.072 – Aviation Computer Emergency Response Team (AV-CERT).

Rulemaking Programme for 2017-2021 makes a fundamental point as it identifies the relationship between safety and security within the context of cybersecurity. This is an important ideological change from the 2016-2020 programme, which focused on 'unlawful interference'. Whereas 'unlawful interference' is terminology predominantly found within the context of security,[121] the 2017 programme begins to link cybersecurity with safety. This is further supported by the programme, which states that the "safety actions in this area are aimed towards mitigating the risks posed by cybersecurity. The impact of security in safety is a strategic priority".[122] It must, however, be stressed that safety and security are generally regulated in different sources of law, with different focuses and require different measures for implementation. Due to this, it should not be assumed that all cybersecurity regulations are equal.[123]

The Rulemaking Programme 2018-2022 contains four rulemaking tasks.[124]

1. RMT.0648 – Aircraft Cybersecurity.

120 EASA, 'Rulemaking and Safety Promotion Programme Including the European Plan for Aviation Safety (EPAS) 2017-2021', Final, 24 January 2017, at 11. *See*, EASA, 'The European Plan for Aviation Safety: Strategic Priorities', https://www.easa.europa.eu/sites/default/files/dfu/The%20European%20Plan%20for%20Aviation%20Safety%20Strategic%20Priorities%20EPAS.pdf.

121 *See*, Chapter II.10 and Chapter III.

122 EASA, 'Rulemaking and Safety Promotion Programme Including the European Plan for Aviation Safety (EPAS) 2017-2021', Final, 24 January 2017, at 57.

123 *See*, Chapter II.9.3.

124 EASA, 'European Plan for Aviation Safety (EPAS) 2018-2022', Final, 14 November 2017.

2. RMT.0720 – Cybersecurity Risks.
3. SPT.071 – Strategy for Cybersecurity in Aviation.
4. RES.012 – Cybersecurity: Common Aeronautical Vulnerabilities Database.

While SPT.071 appears in both the 2016 and 2017 Rulemaking Programme, it is important to note that the name and content are different. SPT.071 will then be treated as two separate rulemaking tasks. Further, while SPT.072 was included in earlier programmes, due to its implementation, it was not included in the current versions of the programme. The Rulemaking Programme 2019-2023 contains the same four rulemaking tasks as the 2018-2022 document.[125] Therefore, there has been a total of six cybersecurity related rulemaking tasks.

7.2.1 RMT.0648 – Amendments to Certification Specifications

Rulemaking Task RMT.0648 is founded on the following assessment.

> All recently designed large aeroplanes are known to be potentially sensitive to those airworthiness-related security threats due to the interconnectivity features of some of their avionics systems. Moreover, some recent avionics modifications may also render legacy aeroplanes sensitive to this risk.[126]

Therefore, the objective of this Rulemaking Task is "to mitigate the safety effects stemming from cybersecurity risks due to acts of unlawful interference with the aircraft on board electronic networks and systems".[127] This is to be achieved by amending the CSs, notably CS-25 and CS-29 to contain necessary content on cybersecurity. AMCs may also be amended.

On 17 May 2016, the Agency published its ToR for RMT.0648,[128] which set out how it plans to fulfil the objectives set out in the Rulemaking Programme. It sets forth the Agency's activities in this area.

> To review the special condition 'Security Assurance Process to isolate or protect the Aircraft Systems and Networks from internal and external Security Threats'.

---

125 EASA, 'Rulemaking and Safety Promotion Programme Including the European Plan for Aviation Safety (EPAS) 2019-2023', Final, 22 November 2018.

126 EASA, 'Terms of Reference for a Rulemaking Task Aircraft Cybersecurity', RMT.0648 – Issue 1, 17 May 2016, at 2.

127 EASA, 'Rulemaking Programme 2016-2020', Final, 11 December 2015, at 33.

128 EASA, 'Terms of Reference for a Rulemaking Task Aircraft Cybersecurity', RMT.0648 – Issue 1, 17 May 2016.

> To take into account the recommendations of the [Aircraft Systems Information Security/Protection] working group.
>
> To propose new certifications specifications (and/or acceptable means of compliance/guidance material) for CS-25 based on analysis of impact (assessment of safe and cost-efficient requirements to prevent the risk).[129]

Cybersecurity has been, up until now, been dealt with under Special Conditions with regards to initial airworthiness and via Instructions for continuing airworthiness.

Amending CS-25, which covers the certification specifications for large aeroplanes,[130] will be the main objective. The publication of an NPA with a proposal to amend CS-25 was scheduled for the first quarter of 2019, with the Executive Director Decision expected in the third quarter of 2019, which will introduce the amendment to CS-25. NPA 2019-01 Aviation Cybersecurity was then published on 22 February 2019:[131]

> The objective of this Notice of Proposed Amendment (NPA) is to mitigate the potential effects of cybersecurity threats on safety. Such threats could be the consequences of intentional unauthorised acts of interference with aircraft on-board electronic networks and systems.
>
> This NPA proposes amendments to CS-23, CS-25, CS-27, CS-29, CS-E, CS-ETSO, CS-P, and, as applicable to their related acceptable means of compliance (AMC)/guidance material (GM), together with AMC-20. The amendments would introduce cybersecurity provisions into the relevant certification specifications (CSs), taking into account the existing special conditions (SCs) and the recommendations of the Aviation Rulemaking Advisory Committee (ARAC) regarding aircraft systems information security/protection (ASISP).
>
> The proposed amendments are expected to contribute to updating the EASA CSs to reflect the state of the art of protection of products and equipment against cybersecurity threats. They are also expected to improve harmonisation with the Federal Aviation Administration (FAA) regulations. Overall, they

---

129 EASA, 'Terms of Reference for a Rulemaking Task Aircraft Cybersecurity', RMT.0648 – Issue 1, 17 May 2016, at 3.

130 Decision No. 2003/2/RM of the Executive Director of the Agency of 17 October 2003 on Certification Specifications, Including Airworthiness Codes and Acceptable Means of Compliance, for Large Aeroplanes ("CS-25»).

131 EASA, Notice of Proposed Amendment 2019-01 – Aircraft Cybersecurity: https://www.easa.europa.eu/sites/default/files/dfu/NPA%202019-01.pdf.

would improve safety, would have neither social nor environmental impact, and would have a neutral-to-positive economic impact.

Stakeholder comments were accepted until May 2019, whereby they would be reviewed, and the amended document would be turned into an Opinion designed for the Commission to adopt as a Regulation.

This task will also consider other certification specifications: CS-23,[132] CS-27,[133] CS-29,[134] CS-E,[135] CS-ETSO[136] and CS-P.[137] This could either be done in parallel to the work on CS-25 or afterwards under a separate rulemaking task. Therefore, it can be expected that the Agency's 'soft law', where relevant to cybersecurity, will be amended as required. Consideration must also be given to how changes to 'soft law', such as CSs, can impact EASA's external relations, notably in the context of its Bilateral Aviation Safety Agreements which concern "mutual acceptance of findings of compliance by Technical Agents or Aviation Authorities between the EU and a Third Country".[138]

This rulemaking task is centred on aviation certification, however, certification is not confined to EASA, as different initiatives have taken place both nationally and internationally within the context of cybersecurity.

1. International certification schemes.
   1.1 Common Criteria Certification – ISO 15408.
   1.2 The ISASecure Certification Programme.
   1.3 Information Technology Security Evaluation Criteria.
   1.4 Federal Information Processing Standards 140.
2. National Certification schemes.
   2.1 French.

132 Decision No. 2003/14/RM of the Executive Director of the Agency of 14 November 2003 on Certification Specifications, Including Airworthiness Codes and Acceptable Means of Compliance for Normal, Utility, Aerobatic and Commuter Category Aeroplanes («CS-23»).

133 Decision No. 2003/15/RM of the Executive Director of the Agency of 14 November 2003 on Certification Specifications for Small Rotorcraft («CS-27»).

134 Decision No. 2003/16/RM of the Executive Director of the Agency of 14 November 2003 on Certification Specifications for Large Rotorcraft («CS-29»).

135 Decision No. 2003/9/RM of the Executive Director of the Agency of 24 October 2003 on Certification Specifications, Including Airworthiness Codes and Acceptable Means of Compliance, for Engines («CS-E»).

136 Decision No. 2003/10/RM of the Executive Director of the Agency of 24 October 2003 on Certification Specifications, Including Airworthiness Codes and Acceptable Means of Compliance, for European Technical Standard Orders («CS-ETSO»).

137 Decision No. 2003/7/RM of the Executive Director of the Agency of 24 October 2003 on Certification Specifications, Including Airworthiness Codes and Acceptable Means of Compliance, for Propellers («CS-P»).

138 P. Mendes de Leon, 2017, at 344-345.

2.2 German.
2.3 UK.
3. Other initiatives.
3.1 Industrial Automation and Control Systems.
3.2 Common Criteria Recognition Arrangement.
3.3 Senior Officials Group – Information Systems Security.
3.4 UL 2900 Certification.
3.5 Secure Change.
3.6 EN50128.
3.7 IEC61508.
3.8 ISO 27001/27002.[139]

Certification is deemed important as it establishes and preserves "trust and security, ICT products and services need to directly incorporate security features in the early stages of their technical design and development (security by design)".[140] Furthermore, customers and users are "able to ascertain the level of security assurance of the products and services they procure or purchase".[141] Aviation stakeholders may then rely on these general cybersecurity certification schemes due to the absence of aviation specific ones and EASA may utilise them to develop their own standards.

The work of standardisation bodies, notably within the context of aviation, should not be overlooked. This is because they are helping with harmonisation while regulations are being developed, they are being developed by industry so they meet the needs of those developing and using aviation technologies, and they can provide flexibility and can be amended more quickly than the law so there is a level of adaptability. The work of the European Cyber Security for Aviation Standards Coordination Group (ECSCG), which is under EUROCAE, will serve as an example of standardisation efforts within the context of cybersecurity, whereby it is explicitly referenced in NPA 2019-01. The ECSCG "is a joint coordination and advisory group established to coordinate the cyber security for aviation related standardisation activities. A specific focus will be on activities stemming from the EC and EASA regulations, this does not exclude other market-driven standards".[142] Other important standard bodies, of which some are referenced by EASA in

139 This is a non-exhaustive list. *See*, Commission, Commission Staff Working Document, Impact Assessment, Accompanying the Document, Proposal for a Regulation of the European Parliament and of the Council on ENISA, the 'EU Cybersecurity Agency', and Repealing Regulation (EU) 526/2013, and on Information and Communication Technology Cybersecurity Certification ('Cybersecurity Act'), Brussels, Belgium, SWD (2017) 500 Final Part 4/6, 13 September 2017 at 38-41.

140 Proposal, COM (2017) 477 Final, 2017/0225 (COD), 13 September 2017, at 8.

141 *Ibid.*

142 EUROCAE, 'The European Cyber security for aviation Standards Coordination Group (ECSCG)', https://www.eurocae.net/about-us/ecscg/.

NPA 2019-01, include ASD-STAN, which is an Associated Body to the European Committee for Standardization for Aerospace Standards, the Radio Technical Commission for Aeronautics, the National Institute of Standards and Technology and the International Organization for Standardization.

On 13 September 2017, the Commission issued a proposal for a regulation on ENISA[143] and on ICT cybersecurity certification which would introduce an EU cybersecurity certification framework. The Regulation, per Article 1(b) of the Proposal, will lay "down a framework for the establishment of European cybersecurity certification schemes for the purpose of ensuring an adequate level of cybersecurity of ICT products and services in the Union. Such framework shall apply without prejudice to specific provisions regarding voluntary or mandatory certification in other Union acts". Pursuant to Article 2(9) of the Proposal, the European Cybersecurity Certification Scheme is "the comprehensive set of rules, technical requirements, standards and procedures defined at Union level applying to the certification of Information and Communication Technology (ICT) products and services falling under the scope of that specific scheme". The Scheme will be "prepared by ENISA, with the assistance, expert advice and close cooperation of the European Cybersecurity Certification Group (see below), and adopted by the Commission by means of implementing acts".[144] Once adopted, "manufacturers of ICT products or providers of ICT services will be able to submit an application for certification of their products or services to a conformity assessment body of their choice. Conformity assessment bodies should be accredited by an accreditation body if they comply with certain specified requirements".[145] Certification will be on a voluntary basis unless it is specified in EU law. Therefore, aviation manufactures of ICT products will be able to apply for certifica-

143 *See*, Chapter II.6.3.
144 Proposal, COM (2017) 477 Final, 2017/0225 (COD), 13 September 2017, at 10.
145 *Ibid*., at 11.

tion.[146] It is unclear whether aviation ICT products will be mandated in EU law to acquire a cybersecurity certificate. The Recital provides a clue as it makes a link between the proposed regulation and aviation.

> (30) To ensure that it fully achieves its objectives, the Agency should liaise with relevant institutions, agencies and bodies, including CERT-EU, European Cybercrime Centre (EC3) at Europol, European Defence Agency (EDA), European Agency for the Operational Management of Large-Scale IT Systems (EU-LISA), European Aviation Safety Agency (EASA) and any other EU Agency that is involved in cybersecurity.[147]

While it shows a link between the regulation and aviation, it does not go as far as to say aviation will be a part of the mandatory group for cybersecurity certification. However, EASA is aware of this overlap and is keen to avoid duplication of efforts.

---

146 Art. 2(11) of the Proposal states that "'ICT product and service' means any element or group of elements of network and information systems". This is elaborated on in Art. 2(2) of the Proposal, which states that "'network and information system' means a system within the meaning of point (1) of Article 4 of Directive (EU) 2016/1148". Art. 4(1) of NIS Directive declares:

> For the purposes of this Directive, the following definitions apply:
> 1. 'network and information system' means:
>   a. an electronic communications network within the meaning of point (a) of Art. 2 of Directive 2002/21/EC;
>   b. any device or group of interconnected or related devices, one or more of which, pursuant to a program, perform automatic processing of digital data; or
>   c. digital data stored, processed, retrieved or transmitted by elements covered under points (a) and (b) for the purposes of their operation, use, protection and maintenance;

Finally, Art. 2(a) of Directive 2002/21/EC declares:

> (a) 'electronic communications network' means transmission systems and, where applicable, switching or routing equipment and other resources which permit the conveyance of signals by wire, by radio, by optical or by other electromagnetic means, including satellite networks, fixed (circuit- and packet-switched, including Internet) and mobile terrestrial networks, electricity cable systems, to the extent that they are used for the purpose of transmitting signals, networks used for radio and television broadcasting, and cable television networks, irrespective of the type of information conveyed.

Directive No. 2002/21/EC of the European Parliament and of the Council of 7 March 2002 on a Common Regulatory Framework for Electronic Communications Networks and Services (Framework Directive), OJ L 108, 24 April 2002, pp. 33-50.

147 Other EU entities may also be relevant and should not be overlooked. These include:

1. Agency for Law Enforcement Training.
2. Agency for Law Enforcement Cooperation.
3. Joint European Union Diplomatic Response to Malicious Cyber Attacks.
4. European Cybersecurity Research and Competence Centre.

7.2.2 SPT.071 – Cybersecurity Road Map

The second element of the rulemaking programme sets in motion the establishment of a cybersecurity Road Map.

> The concrete actions to be taken will be identified in a road map developed jointly by the European Commission and EASA in close cooperation with EU Member States and industry. This road map should be developed in order to avoid duplication and prevent jeopardising the effort already initiated by the industry. Furthermore, a cybersecurity strategy is being implemented in the EU for the protection of EU citizens against cybercrime. This strategy, together with the EU aviation strategy, will pave the way for a secure and safe air transport system.[148]

This tool will be very useful for determining how the EU aviation legal regime, as well as the other aspects of EASA's activities, will develop within the scope of cybersecurity. The completion date of 2016 was assigned in Rulemaking Programme 2016-2020, which was then extended to 2017 by Rulemaking Programme 2017-2022. The latter deadline was been met, with references being made to the Road Map in Agency literature, but as it was an internal document, comments cannot be made to its exact content. The Agency's cybersecurity activities have surpassed the Road Map and, because of this, it is no longer used.

7.2.3 SPT.072 – Aviation Computer Emergency Response Team

SPT.072 introduces the desire to create an aviation computer emergency response team called 'AV-CERT'. The core goal of AV-CERT would be "to provide information and assistance to European aviation stakeholders in implementing proactive measures to reduce the risks of computer security incidents as well as in responding to such incidents when they occur".[149] Further to this, AV-CERT will "offer support to European aviation stakeholders on the prevention of and response to information-related security incidents". In order to achieve this, it will:

1. raise cybersecurity awareness through cybersecurity promotion initiatives,
2. advise EASA and the Commission on aviation cybersecurity policy and regulatory matters, and
3. advise the Commission, Member States or the European Aviation Crisis Coordination Cell (EACCC) in cases of aviation cybersecurity crisis.

---

148 EASA, 'Rulemaking and Safety Promotion Programme Including the European Plan for Aviation Safety (EPAS) 2017-2021', Final, 24 January 2017, at 57.

149 EASA, European Plan for Aviation Safety 2016-2020, Final, 25 January 2016.

The EACCC was established in May 2010 by the Commission and Eurocontrol. It was set up "to coordinate the management of crisis responses in the European ATM network".[150] Its role is, therefore, "to coordinate the response to those network crisis situations which impact adversely on aviation, in close cooperation with corresponding structures in States".[151] This "includes proposing measures and taking initiatives and, in particular, acquiring and sharing information with the aviation community [...] in a timely manner".[152]

The term 'Computer Emergency Response Team' was first used by the CERT Coordination Center (CERT-CC) at Carnegie Mellon University (CMU).

> Computer Emergency Response Team (CERT): This is a team of experts, either public or private, that manage computer security incidents. The role of a CERT can vary, but it typically involves taking preventative measures, as well as corrective measures.

Its infrastructure is located at the Software Engineering Institute, which is a federally funded research and development centre located at CMU in the US. CERT, as used by CMU, is not an acronym and it has been registered by CMU with the US Patent and Trademark Office, and with other authorities around the world and has an EU Communities Trademark. CMU has enforced its trademark in the past and claims that it investigates all allegations of non-authorised use of its trademark. Entities that aim to improve cybersecurity are able to apply for a licence to use the 'CERT' trademark.

7.2.4 RMT.0720 – Cybersecurity Risks

RMT.0720 presents a solid plan for the creation of specific aviation safety rules for cybersecurity.

> The specific objective of this task is to create a regulatory system which efficiently contributes to the protection of the aviation system from cyber-attacks and their consequences.

The idea is to introduce a Commission regulation covering all domains of aviation safety, including; design, production, maintenance, operations, aircrew, ATM/ANS and aerodromes.

150 Eurocontrol, 'The European Aviation Crisis Coordination Cell (EACCC)', http://www.eurocontrol.int/articles/european-aviation-crisis-coordination-cell-eaccc.

151 *Ibid.*

152 *Ibid.*

Two main approaches have been taken to regulating new areas within aviation. The first involves amending existing rules to reflect the new area. This has been the current approach taken by ICAO within the context of cybersecurity with Amendment 12 to Annex 17.[153] Further, the Annexes of the new EU safety regulation[154] arguably have elements of cybersecurity contained in general aviation safety rules. The second approach involves creating a *sui generis* set of rules that deal with that specific area of law. This approach has, for example, been taken by the EU with regard to unmanned aircraft. The Agency has opted for the latter approach under RMT.0720.

The latter approach was appropriate for unmanned aircraft as there was a specific focus; the design, manufacturing, training, use, oversight and enforcement of that specific type of aircraft. Cybersecurity, on the other hand, presents a less tangible subject as it does not relate to one type of aircraft, one type of activity, one single stakeholder, or one regulatory activity, such as design, manufacturing, training, use, oversight and enforcement. Therefore, creating a *sui generis* regulation will certainly be an ambitious task, and time will tell how successful this endeavour will be.

NPA 2019-07 titled 'Management of Information Security Risks' was published on 27 May 2019. The proposed Regulation sets out to establish the requirements to be met by organisations and competent authorities involved in civil aviation activities in order to identify, protect from, detect, respond to and recover from those information security incidents which could potentially affect aviation safety. These provisions shall apply to competent authorities and organisations in all aviation domains, including:

1. Design.
2. Production.
3. Management of continuing airworthiness.
4. Maintenance.
5. Air operations.
6. Aircrew.
7. Air traffic management/air navigation services (ATM/ANS).
8. Aerodromes.

Organisations involved in activities where the level of risk is acceptable or where there is no cybersecurity risk shall not have to comply with the Regulation. There are 2 Annexes, which contain the legally binding technical requirements for implementation:

1. Annex I – Organisation Requirements.

153 *See*, Chapter III.4.2.
154 *See*, Chapter II.9.

2. Annex II – Authority Requirements.

EASA will also establish AMC and GM. It can be expected that industry will be key to the development of the AMCs and GMs, through cooperation with EASA and working through industry standard bodies. NPA 2019-07 may attract many comments as, for example, it allows the NIS Directive to satisfy some of its requirements. Therefore, a Directive, which is implemented differently in each state, can be used as a tool for applying a Regulation, which is applied equally across all states. Further, EASA's mandate is *aviation safety*, and there is an argument that the Regulation goes beyond this mandate. For example, some of the requirements involve traditional 'security' matters, like mandating criminal record checks, or the requirements are simply procedural or bureaucratic with no safety implication at all. The Opinion is scheduled for the fourth quarter of 2020. The Opinion will then be issued to the Commission for its consideration, with the goal of turning it into a Commission Implementing Regulation that addresses the management of information security risks by civil aviation organisations.

7.2.5 SPT.071 – Strategy for Cybersecurity in Aviation

An aviation cybersecurity strategy will be developed by the Commission and EASA, which will work in close cooperation with EU Member States and industry stakeholders. As set out in SPT.071 – Strategy for Cybersecurity in Aviation, this strategy will include, *inter alia*, six actions.

1. Information sharing.
2. Research and studies.
3. Event investigation and response.
4. Knowledge and competence building.
5. International cooperation and harmonisation.
6. Regulatory activities and development of industry standards.

The strategy will work alongside and in cooperation with the wider cybersecurity strategy being implemented in the EU for the protection of EU citizens against cyberattacks.[155] It is due for publication as a Strategy Paper at some point in 2019.

SPT.071 contains nothing new and merely repeats what was said in the Bucharest Declaration and Krakow Declaration. Therefore, it can be seen as a step towards the implementation of the points raised in the two Declarations.

There is potential overlap between SPT.071 and other activities of the Agency. For example, 'event investigation and response' may simply be a reflection of the work of a

155 *See*, Chapter II.1.

CERT, but it could also extend to core aviation activities of accident and incident reporting under Regulation 996/2010[156] and Regulation 376/2014,[157] as well as Aircraft Accident and Incident Investigation under Annex 13 of the Chicago Convention.[158] For example, Regulation 376/2014 requires the establishment of a mandatory and a voluntary occurrence reporting system, which will collect any event representing a serious risk to an aircraft or its occupants. Alternatively, the strategy may envisage another mechanism for event investigation and response. It is, thus, unclear how much the strategy will overlap with other efforts by the Agency in these areas.

7.2.6 RES.012 – Common Aeronautical Vulnerabilities Database

The final task pertains to the development of a 'vulnerability database' which would involve collecting, maintaining and disseminating information about discovered vulnerabilities found in major transport information systems. Little information is given about this task, except that it is due in 2021. Therefore, the scope of this task is unclear as it is not known who the Agency will collect the information from and to whom it will circulate to. Furthermore, it is unclear how this database will differentiate from and react with the Agency's other cooperation activities, such as the ECCSA.

8 Information Sharing and Cooperation in the European Union

8.1 Foundations of the European Centre for Cybersecurity in Aviation

In order to facilitate work with third parties, the Agency often enters into a Memorandum of Cooperation (MoC) or a Memorandum of Understanding (MoU).

> Memorandum of Cooperation (MoC): A non-binding document between two or more parties that outlines the key areas on the cooperation.
>
> Memorandum of Understanding (MoU): A non-binding document between two or more parties that expresses a common will and indicates the intended common line of action.

156 Regulation (EU) No. 996/2010.

157 Regulation (EU) No. 376/2014 of the European Parliament and of the Council of 3 April 2014 on the Reporting, Analysis and Follow-Up of Occurrences in Civil Aviation, Amending Regulation (EU) No. 996/2010 of the European Parliament and of the Council and Repealing Directive No. 2003/42/EC of the European Parliament and of the Council and Commission Regulations (EC) No. 1321/2007 and (EC) No. 1330/2007, OJ L 122, 24 April 2014, pp. 18-43, Art. 1(1).

158 Ann. 13 to the Chicago Convention, Aircraft Accident and Incident Investigation, Eleventh Edition, July 2016.

In light of this, on 10 February 2017, EASA signed an MoC with CERT-EU. The MoC sets out the foundations of cooperation between EASA and CERT-EU as they will work together to establish the ECCSA.[159] The mission of the ECCSA is

> to provide information and assistance to European aviation manufacturers, airlines, maintenance organizations, air navigation service providers, etc. in order to protect the critical elements of the system such as aircraft, navigation and surveillance systems, datalinks, airports, etc. ECCSE will cover the full spectrum of aviation.[160]

EASA's role in this cooperation is important as it has specialised technical aviation knowledge and is a trusted member of the aviation community. Therefore, it is well positioned to act as a bridge between the aviation community and the cybersecurity community.

8.2 *Computer Emergency Response Team – European Union*

CERT-EU was established by the EU on 11 September 2012 to act as the CERT for EU Institutions, bodies and agencies. Members of CERT-EU comprises IT security experts from the main EU Institutions; Commission, General Secretariat of the Council, European Parliament, Committee of the Regions, and Economic and Social Committee. The role of CERT-EU is to improve the EU's response to cyberattacks, which covers all control modes of prevention, detection and corrective.[161] It also works in close cooperation with Member States' CERTs, as well as non-Member States and with IT security companies.

CERT-EU, therefore, has specialised cybersecurity knowledge and will, within the cooperation "make available cybersecurity tools and information for the development of ECCSA's specialized aviation cybersecurity components and services".[162] Therefore, EASA will benefit from the general expertise provided by CERT-EU and CERT-EU will benefit from the specialised aviation knowledge provided by EASA, which will allow the ECCSA to better cover the overall EU threat landscape.

159 EASA, 'Implementation of a European Centre for Cyber Security in Aviation (ECCSA)', On Air, Issue 14: Cybersecurity in Aviation, 4 April 2017, https://www.easa.europa.eu/newsroom-and-events/news/implementation-european-centre-cyber-security-aviationeccsa.

160 EASA, 'EASA Cooperate with CERT-EU on Cybersecurity', 14 February 2017, https://www.easa.europa.eu/newsroom-and-events/press-releases/easa-cooperate-cert-eu-cybersecurity.

161 *See*, Conclusion 3.3.

162 EASA, 'EASA Cooperate with CERT-EU on Cybersecurity', 14 February 2017, https://www.easa.europa.eu/newsroom-and-events/press-releases/easa-cooperate-cert-eu-cybersecurity.

8.3 *Membership of the European Centre for Cybersecurity in Aviation*

Membership of the ECCSA is open to actors of the European aviation system, is on a voluntary basis and allows them to benefit from "intelligence from aviation related sources which will provide awareness of cyberattacks and, on demand, operational means to face cyber security threats".[163] This would include:

1. Aviation manufacturers: Aircraft, avionics systems or ground systems, etc.
2. Aviation organisations: Airlines, maintenance repair organisations, aviation communication services providers, etc.
3. Air Navigation Service Providers (ANSP): National and cross-border ANSPs.[164]

It is desired to make membership inclusive, in order to benefit from numerous aviation stakeholders' expertise. It is open to all European aviation stakeholders to apply; however, acceptance is not guaranteed and is conditional upon applicable security criteria.

The activities of ECCSA are likely to remain restricted between its members, with only occasional general statements being made so to make the public aware of the work being carried out in this area.

It was foreseen that the first stage in implementing the ECCSA, between 2017 and 2018, will include:

1. A public website reporting cyber security news and ECCSA initiatives,
2. Open Source Intelligence services for members,
3. A collaboration platform for members to exchange sectorial cyber security information.[165]

The issues with information sharing groups, as presented in Chapter II.5, also persists for the ECCSA.

---

163 *Ibid.*
164 *Ibid.*
165 *Ibid.*

8.4 *Network and Information Systems Directive*

The NIS Directive was adopted by the European Parliament on 6 July 2016 and entered into force in August 2016.

> Network and Information Security: "[T]he ability of a network or an information system to resist, at a given level of confidence, accidental events or malicious actions that compromise the availability, authenticity, integrity and confidentiality of stored or transmitted data and the related services offered by or accessible via these networks and systems".[166]

Member States then had 21 months to transpose the Directive into national laws and six months more to identify operators of essential services.[167] Pursuant to Article 1 of the NIS Directive, the subject and scope are as follows:

1. This Directive lays down measures with a view to achieving a high common level of security of network and information systems within the Union so as to improve the functioning of the internal market.
2. To that end, this Directive:
   a. lays down obligations for all Member States to adopt a national strategy on the security of network and information systems;
   b. creates a Cooperation Group in order to support and facilitate strategic cooperation and the exchange of information among Member States and to develop trust and confidence amongst them;
   c. creates a computer security incident response teams network ('CSIRTs network') in order to contribute to the development of trust and confidence between Member States and to promote swift and effective operational cooperation;
   d. establishes security and notification requirements for operators of essential services and for digital service providers;
   e. lays down obligations for Member States to designate national competent authorities, single points of contact and CSIRTs with tasks related to the security of network and information systems.

166 Commission, Communication from the Commission to the Council, the European Parliament, the European Economic and Social Committee and the Committee of the Regions: Network and Information Security: Proposal for A European Policy Approach, Brussels, 6 June 2001, COM(2001)298 Final.
167 Directive (EU) No. 2016/1148, Ch. VII.

The Directive applies to two separate subjects.

1. Operators of Essential Services that are established in the EU, and
2. Digital Service Providers that offer services to persons within the EU.

Pursuant to Article 4(4), an operator of essential services shall mean "a public or private entity of a type referred to in Annex II, which meets the criteria laid down in Article 5 (2)". Following this, the operators of essential services shall be as follows:

a. an entity provides a service which is essential for the maintenance of critical societal and/or economic activities;
b. the provision of that service depends on network and information systems; and
c. an incident would have significant disruptive effects on the provision of that service.

This covers any sector and subsector referred to in Annex II.

1. Energy.
2. Transport.
3. Banking.
4. Financial market infrastructures.
5. Healthcare.
6. Drinking water supply and infrastructure.
7. Digital infrastructure.

Annex II.2 then lists which stakeholders shall be included as Operators of Essential Services in air transport.

> Air carriers as defined in point (4) of Article 3 of Regulation (EC) No. 300/2008 of the European Parliament and of the Council.[168]

168 Regulation 300/2008, Art. 3(4): 'air carrier' means an air transport undertaking holding a valid operating licence or equivalent.

> Airport managing bodies as defined in point (2) of Article 2 of Directive 2009/12/EC of the European Parliament and of the Council, airports as defined in point (1) of Article 2 of that Directive, including the core airports listed in Section 2 of Annex II to Regulation (EU) No. 1315/2013 of the European Parliament and of the Council, and entities operating ancillary installations contained within airports.[169]
>
> Traffic management control operators providing air traffic control (ATC) services as defined in point (1) of Article 2 of Regulation (EC) No. 549/2004 of the European Parliament and of the Council.[170]

Pursuant to Article 4(6), the term "'digital service provider' means any legal person that provides a digital service". This translates into, as specified in Annex III, as search engines, cloud computing services or online marketplaces. As a result, Member States will have to act within the transport sector. This includes:

1. Article 7 – National strategy on the security of network and information systems.
2. Article 8 – National competent authorities and single point of contact.
3. Article 9 – Computer security incident response teams (CSIRTs).
4. Article 10 – Cooperation at national level.
5. Article 14 – Security requirements and incident notification.
6. Article 15 – Implementation and enforcement.

---

169 Directive No. 2009/12/EC of the European Parliament and of the Council of 11 March 2009 on Airport Charges, OJ L 70, 14 March 2009, pp. 11-16, Art 2:

1. 'airport' means any land area specifically adapted for the landing, taking-off and manoeuvring of aircraft, including the ancillary installations which these operations may involve for the requirements of aircraft traffic and services, including the installations needed to assist commercial air services.
2. 'airport managing body' means a body which, in conjunction with other activities or not as the case may be, has as its objective under national laws, regulations or contracts the administration and management of the airport or airport network infrastructures and the coordination and control of the activities of the different operators present in the airports or airport network concerned.

Regulation (EU) No. 1315/2013 of the European Parliament and of the Council of 11 December 2013 on Union Guidelines for the Development of the Trans-European Transport Network and Repealing Decision No. 661/2010/EU, OJ L 348, 20 December 2013, pp. 1-128, Ann. II.2.

170 Regulation (EC) No. 549/2004 of the European Parliament and of the Council of 10 March 2004 Laying Down the Framework for the Creation of the Single European Sky (The Framework Regulation) – Statement by the Member States on Military Issues Related to the Single European Sky, OJ L 96, 31 March 2004, pp. 1-9, Art 2:

1. 'air traffic control (ATC) service' means a service provided for the purpose of: (a) preventing collisions: – between aircraft, and – in the manoeuvring area between aircraft and obstructions; and (b) expediting and maintaining an orderly flow of air traffic.

There is, therefore, a relationship between the NIS Directive and Member State with regard to CSIRTs, cooperation and notification. As a Directive, Member States must implement it at a national level, so the ECCSA may be deemed as satisfaction of Article 9 for air transport by a Member State. This is supported by Article 1(7) of the NIS Directive:

> Where a sector-specific Union legal act requires operators of essential services or digital service providers either to ensure the security of their network and information systems or to notify incidents, provided that such requirements are at least equivalent in effect to the obligations laid down in this Directive, those provisions of that sector-specific Union legal act shall apply.

Therefore, the *lex specialis* – i.e. aviation – can be used in place of the NIS Directive, so the ECCSA *could* become the sector-specific CERT for aviation. As the ECCSA has a wider scope than a CERT under the NIS Directive, such as the inclusion on non-EU stakeholders and stakeholders engaged in non-essential services, this could benefit aviation cybersecurity. There is, thus, an interconnection between the NIS Directive and the aviation safety regime, and this should be taken into consideration and the exact relationship should be determined.

8.5 'Large-Scale' Cybersecurity Incidents and Crises

It must be highlighted that the NIS Directive does not provide for a Union cooperation framework for 'large-scale' cybersecurity incidents and crises. As a result, the EU Member States requested the Commission to prepare a Blueprint for consideration by the EU bodies and other relevant stakeholders. The Blueprint is an Annex to Commission Recommendation 2017/1584 on coordinated response to large-scale cybersecurity incidents and crises.[171] The Recommendation states that "Member States and EU institutions should establish an EU Cybersecurity Crisis Response Framework integrating the objectives and modalities of cooperation presented in the Blueprint following the guiding principles described therein".[172] This is limited to "cybersecurity incidents which cause disruption too extensive for a concerned Member State to handle on its own or which

171 Commission Recommendation (EU) No. 2017/1584 of 13 September 2017 on Coordinated Response to Large-Scale Cybersecurity Incidents and Crises, C/2017/6100, OJ L 239, 19 September 2017, pp. 36-58.

172 *Ibid.*, at Recommendation 1. *See*, for more information, Commission, Annex to the Commission Recommendation on Coordinated Response to Large Scale Cybersecurity Incidents and Crises, 13 September 2017, C (2017) 6100 Final Ann. 1; and Opinion of the European Economic and Social Committee on the 'Communication from the Commission to the European Parliament, the Council, the European Economic and Social Committee and the Committee of the Regions on Critical Information Infrastructure Protection "Protecting Europe from Large Scale Cyber-Attacks and Disruptions: Enhancing Preparedness, Security and Resilience"', 22 September 2010, COM (2009) 149 Final, OJ C 255, pp. 98-102.

affect two or more Member States or EU institutions with such a wide-ranging and significant impact of technical or political significance that they require timely policy coordination and response at Union political level. Such large-scale cybersecurity incidents are considered a cybersecurity 'crisis'".[173] Four general guiding principles apply:
1. Proportionality.
2. Subsidiarity.[174]
3. Complementarity.
4. Confidentiality of information.

The core objectives of cooperation under the Blueprint shall consider political, operational and technical aspects, and shall:
1. enable an effective response,
2. share situational awareness, and
3. agree on key public communication messages.

These are then explained in detail throughout the rest of the Annex and Appendix. Therefore, aviation may fall under this Regulation, when the scale is of sufficient size. Thus, this is a further example of how non-aviation EU law may have an impact on aviation stakeholders.

8.6 Responsible Disclosure

IOActive, which is a White Hat cybersecurity company, found vulnerabilities in the on board passenger WiFi provided through satellites. IOActive informed EASA of this through 'responsible disclosure', which is informing the relevant stakeholders prior to the vulnerability being made public.[175] Once EASA was made aware, it coordinated with operators and other authorities, and then performed independent verifications.[176] As IOActive was only able to access passengers' phones, laptops and tablets, there was no risk to safety as access could not be gained to the aircraft. The vulnerability was then closed. This example shows the benefit of cross-sector cooperation, aviation stakeholder cooperation, the use of White Hats in improving cybersecurity, and the need to continually monitor old and new technology. This example should hopefully encourage over White Hats to engage in responsible disclosure and confirm EASA's role in this process, of which it should actively promote.

173 *Ibid.*, at Annex – Introduction.
174 *See*, Regulation 2018/1139, Recital 88.
175 EASA, 'EASA Engaged in a 'Responsible Disclosure' of Cybersecurity Issues in Coordination with Operators and Authorities', https://www.easa.europa.eu/newsroom-and-events/news/easa-engaged-%E2%80%9Cresponsible-disclosure%E2%80%9D-cybersecurity-issues-coordination.
176 *Ibid.*

9 New Aviation Safety Regulation

9.1 European Union Aviation Strategy

In 2015, the Commission published its Aviation Strategy to Enhance the Competitiveness of the EU Aviation Sector.[177] The strategy stated that the EU wanted to repeal and replace Regulation 216/2008. The objective of such endeavour was to produce a Parliament and Council aviation safety Regulation that would build "upon the lessons learnt and experiences gained throughout the life of EASA that can effectively regulate aviation safety for the next decade".[178] One of the goals was to introduce new specific areas to the EU aviation safety regime.

> Under the present initiative a limited number of specific areas are proposed to be added to this overall Union aviation safety framework, namely unmanned aircraft, safety of ground handling services and security aspects of aircraft and aviation systems' design, including cybersecurity.[179]

It is argued, as evidenced above in Chapter II.2, that the new Regulation would not add cybersecurity to the scope of the aviation safety regime found under Regulation 216/2008. Rather, it represents the first time that the term 'cybersecurity' features directly in the EU aviation safety cover Regulation. As a result, this merely positively codifies the link between aviation safety and cybersecurity within a binding legal context. The new Regulation is Regulation 2018/1139 of the European Parliament and of the Council of 4 July 2018 on Common Rules in the Field of Civil Aviation and Establishing a European Union Aviation Safety Agency.

9.2 Objective and Scope of the European Aviation Safety Agency

Pursuant to Article 1(1) of Regulation 2018/1139, the main objective is to "maintain a high uniform level of civil aviation safety in the Union". In addition to this, Article 1(2) (a)-(k) then sets forth 11 additional objectives.

177 Commission, Commission Presents a New Aviation Strategy for Europe, Brussels, Press Release, IP/15/6144, 7 December 2015.
178 P. Mendes de Leon, 2017, at 347.
179 Proposal, COM/2015/0613 Final – 2015/0277 (COD), 7 December 2015, at 4.

a. contribute to the wider Union aviation policy and to the improvement of the overall performance of the civil aviation sector;
b. facilitate, in the fields covered by this Regulation, the free movement of goods, persons, services and capital, providing a level playing field for all actors in the internal aviation market, and improve the competitiveness of the Union's aviation industry;
c. contribute to a high, uniform level of environmental protection;
d. facilitate, in the fields covered by this Regulation, the movement of goods, services and personnel worldwide, by establishing appropriate cooperation with third countries and their aviation authorities, and by promoting the mutual acceptance of certificates and other relevant documents;
e. promote cost-efficiency, by, inter alia, avoiding duplication, and promoting effectiveness in regulatory, certification and oversight processes as well as an efficient use of related resources at Union and national level;
f. contribute, in the fields covered by this Regulation, to establishing and maintaining a high uniform level of civil aviation security;
g. assist Member States, in the fields covered by this Regulation, in exercising their rights and fulfilling their obligations under the Chicago Convention, by ensuring a common interpretation and a uniform and timely implementation of its provisions, as appropriate;
h. promote, worldwide, the views of the Union regarding civil aviation standards and civil aviation rules, by establishing appropriate cooperation with third countries and international organisations;
i. promote research and innovation, inter alia, in regulatory, certification and oversight processes;
j. promote, in the fields covered by this Regulation, technical and operational interoperability and the sharing of administrative best practices;
k. support passenger confidence in a safe civil aviation.

The objectives prescribed in Article 1(1) and (2) shall be achieved pursuant to Article 1 (3).

The objectives set out in paragraphs 1 and 2 shall be achieved by, inter alia:
a. the preparation, adoption and uniform application of all necessary acts;
b. the taking of measures to improve safety standards;
c. ensuring that the declarations and certificates issued in accordance with this Regulation, and with the delegated and implementing acts adopted on the basis thereof, are valid and recognised throughout the Union, without any additional requirements;

d. the development, with the involvement of standardisation and other industry bodies, of detailed technical standards to be used as a means of compliance with this Regulation, and with the delegated and implementing acts adopted on the basis thereof, where appropriate;
e. the establishment of an independent European Union Aviation Safety Agency (the 'Agency');
f. the uniform implementation of all necessary acts by the national competent authorities and the Agency, within their respective areas of responsibility;
g. the gathering, analysis and exchange of information to support evidence-based decision making;
h. the undertaking of awareness and promotion initiatives, including training, communication and dissemination of relevant information.

Consequently, Regulation 2018/1139 provides competencies to the Agency to handle safety related cybersecurity matters in line with Article 1.

While having a broad scope of competencies, the Agency is limited by Regulation 2018/1139. First, the Regulation only applies to *civil* activities per Article 2(3)(a)-(c).

3 This Regulation shall not apply to:
   a. aircraft, and their engines, propellers, parts, non-installed equipment and equipment to control aircraft remotely, while carrying out military, customs, police, search and rescue, firefighting, border control, coastguard or similar activities or services under the control and responsibility of a Member State, undertaken in the public interest by or on behalf of a body vested with the powers of a public authority, and the personnel and organisations involved in the activities and services performed by those aircraft;
   b. aerodromes or parts thereof, as well as equipment, personnel and organisations, that are controlled and operated by the military;
   c. ATM/ANS, including systems and constituents, personnel and organisations, that are provided or made available by the military.

Second, the Regulation does not apply to:

> the design, production, maintenance and operation of aircraft the operation of which involves low risk for aviation safety, as listed in Annex I, and to the personnel and organisations involved therein, unless the aircraft has been is-

> sued, or has been deemed to have been issued, with a certificate in accordance with Regulation (EC) No. 216/2008.[180]

Annex I of Regulation 2018/1139 also contains a list of aircraft types that are deemed outside the scope of the Regulation and would instead be regulated at a national level by the Member States. This list has been taken from Annex II of Regulation 216/2008 and amended, with the major change being the removal of the 150 kg limitation for unmanned aircraft. Thus, these general limitations must be considered when dealing with cybersecurity matters.

Many of the fundamentals between Regulation 216/2008 and Regulation 2018/1139 remain the same, however, many changes will be brought in by this new Regulation. Here, focus will be given to the content that relates directly and indirectly to cybersecurity.

9.3 Recital to Regulation 2018/1139

9.3.1 Recital 12

In order to understand the context that cybersecurity has within the EU aviation safety legal regime, recourse to the Recitals is required. The first reference to 'cybersecurity' in Regulation 2018/1139 is in Recital 12.

> The measures taken in accordance with this Regulation to regulate civil aviation in the Union, and the delegated and implementing acts adopted on the basis thereof, should correspond and be proportionate to the nature and risks associated with the different types of aircraft, operations and activities they address. Such measures should also, in as far as possible, be formulated in a manner which focuses on objectives to be achieved, while allowing different means of achieving those objectives, and should also foster a systemic approach to civil aviation, taking into account interdependencies between safety and other technical domains of aviation regulation, including cyber security. This should contribute to a more cost-efficient achievement of required safety levels and to the stimulation of technical and operational innovation. Use should be made of recognised industry standards and practices, where it has been found that they ensure compliance with the essential requirements set out in this Regulation.

180 Regulation 2018/1139, Art. 2(3)(d).

Measures taken in accordance with the Regulation shall take into account the relationship between safety and cybersecurity. Recital 12 makes it explicitly clear that the two are distinct areas; cybersecurity is not *stricto sensu* a part of safety.

9.3.2 Recital 59

The distinction is repeated in Recital 59.

> Having regard to the existing interdependencies between safety and security in civil aviation, the Agency should take part in the cooperation concerning the area of aviation security, including cyber-security. It should contribute its expertise to the implementation, by the Commission and by Member States, of Union rules in that area.

Recital 59 takes the classification of 'cybersecurity' further than Recital 12 as it places cybersecurity under the scope of aviation security.

9.3.3 Use of the Terms 'Security' and 'Safety'

The terms 'security' and 'safety' in aviation can lead to some confusion as they often go undefined in EU aviation law, as well as in the international law cited in Chapter III, or have ambiguous borders in practice that can result in their use overlapping.

One of the causes behind this issue is that not all languages make a distinction between 'security' and 'safety', thus they only use one term to encapsulate the two words found in the English language. Within the EU, aviation law frequently uses the two terms in English and there is often a demarcation criterion for the scope of the legal text. For example, Regulation 300/2008, which is on the common rules in the field of civil aviation security, is concerned with aviation security matters.[181]

> This Regulation establishes common rules to protect civil aviation against acts of unlawful interference that jeopardise the security of civil aviation. It also provides the basis for a common interpretation of Annex 17 to the Chicago Convention on International Civil Aviation.[182]

On the contrary, Regulation 216/2008 had the aim per Article 2(1) of establishing and maintaining "a high uniform level of civil aviation safety in Europe". The scope of the

---

181 *See*, Chapter II.10.

182 Regulation (EC) No. 300/2008, Art. 1(1). *See*, for more information, Ann. 17 to the Chicago Convention, Security: Safeguarding International Civil Aviation Against Acts of Unlawful Interference, Tenth Edition (Revised), April 2017.

prior Regulation is overtly stated as 'security' and that of the latter clearly categorised as 'safety'. What this means in practice, from a simple reading of the cited Articles, is somewhat unclear, but what is apparent is that the terms are not used as synonyms, but rather in a distinct manner.

The consequences of this confusion have so far been minimal as each category often exists in its own silo. This is not always the case, as the two can overlap as seen with the Germanwings 9525 crash which had elements of hijacking, which is traditionally treated as a security issue,[183] and links to the pilot's medical fitness and the cockpit door design, which is traditionally handled under the safety regime.[184] The two legal regimes are coming closer together, such as through Regulation 2018/1139 which uses the term 'safety', 'security' and 'cybersecurity'.

Five different languages will be examined to show the different ways that the EU has tried to handle this issue within the context of Article 88 of Regulation 2018/1139.

Table 2 Safety/Security in Different Languages

	Regulation 2018/1139 Article 88	**Regulation 2018/1139 Article 88**	**Regulation 2018/1139 Article 88**	**Regulation 300/2008**	**Regulation 216/2008**
Danish	*Flyvesikkerhed*	*Luftfartssikkerhed* (Security)	*Cybersikkerhed*	*Sikkerhed*	*Sikkerhed*
English	Safety	Security	Cybersecurity	Security	Safety
French	*Sécurité*	*Sûreté*	*Cybersécurité*	*Sûreté*	*Sécurité*
German	*Flugsicherheit*	*Luftsicherheit*	*Cybersicherheit*	*Sicherheit*	*Flug-* and *Sicherheit*
Polish	*Bezpieczeństwo*	*Ochrona*	*Cyberbezpieczeństwo*	*Ochrona*	*Bezpieczeństwo*

It is the case that several of the official EU languages use the same term to express both 'security' and 'safety', such as with Danish which uses the work '*Sikkerhed*'. The Danish text in Article 88, however, actually resorts to English to stress the distinction as it states '*Flyvesikkerhed*' is safety and '*Luftfartssikkerhed* (Security)'; the text utilises English by including 'security' in brackets to make the distinction clearer.

183 *See*, Convention for the Suppression of Unlawful Seizure of Aircraft (The Hague Convention 1970), 16 December 1970, UN Treaty Series 1973 (1970).

184 *See*, Commission Regulation (EU) No. 1178/2011 of 3 November 2011 Laying Down Technical Requirements and Administrative Procedures Related to Civil Aviation Aircrew Pursuant to Regulation (EC) No. 216/2008 of the European Parliament and of the Council, OJ L 311, 25 November 2011, pp. 1-193.

Interestingly, the French version has taken a different approach than that taken at an international level and has opted for two separate terms. For example, the English version of the Preamble to the Chicago Convention uses 'security' and the French version uses '*sécurité*'. The term 'safety', which is much more prevalent in the English version of the Chicago Convention is also '*sécurité*' in the French version. The approach used for Regulation 2018/1139, however, is taken from the different words used in Regulation 300/2008 ('*sûreté*') and Regulation 216/2008 ('*sécurité*'). Therefore, for consistency, the terminology adopted in each Regulation to differentiate 'safety' from 'security' has been adopted in Regulation 2018/1139. Therefore, this context must be understood in order correctly understand the French version of Regulation 2018/1139. For example, the term '*cybersécurité*' is a direct translation into 'cybersecurity', but within the context of Regulation 2018/1139 the translation would be 'cybersafety', which is not a generally accepted term in aviation.[185]

German traditionally uses '*Sicherheit*' to refer to both terms. However, a distinction has been made via a prefix. Firstly, '*Luftsicherheit*' is used to pertain to 'security' and '*Flugsicherheit*' is used for 'safety'. The German version of Regulation 216/2008, however, switches between '*Flugsicherheit*' and '*Sicherheit*' throughout and attaches several suffixes[186] to '*Sicherheit*'. As a result, the use is not clear cut and caution should be exercised when looking at the Regulations and interpreting Table 2.

The Polish version uses two separate words to make the distinction clear. First, it uses '*Bezpieczeństwo*', which in everyday use means both security and safety, to refer to safety only. It then uses '*Ochrona*', which in everyday language means 'protection' to refer to security only. As a result, the translators have had to adopt '*Ochrona*' to make clear the division that is found in the English version. This approach has been generally adopted by previous EU aviation safety legislative instruments,[187] so there is some consistency. However, there is some overlap as, for example, '*Ochrona*' also features in Regulation 216/2008 within the context of environmental protection.

It seems apparent that the drafters intend security and safety to mean different things and use the language available to them to try and distinguish between the two where possible. This study will, therefore, adopt this dichotomy. However, what this shows is that the answer is not based in linguistics, which tries to accommodate the distinction, rather it is found elsewhere in practical usage.

---

185 The use of the term 'cybersafety' is becoming more popular at ICAO. *See*, Chapter III.5.2.2.

186 *See*, as these are not cited, Regulation 216/2008.

187 This refers to Regulation 216/2008 and Regulation 300/2008 which is covered in Chapter II.10.

This conclusion is supported by the various definitions found in EU law and the Annexes to the Chicago Convention. For example, Article 3(2) of Regulation 300/2008 defines 'aviation security' to mean:

> the combination of measures and human and material resources intended to safeguard civil aviation against acts of unlawful interference that jeopardise the security of civil aviation.

Whereas, Annex 17 defines 'security' in its definitions as:

> Safeguarding civil aviation against acts of unlawful interference. This objective is achieved by a combination of measures and human and material resources.[188]

Both definitions show that any measure based on 'security' should attempt to combat 'unlawful interference'.[189] On the other hand, Annex 19 defines 'safety' as:

> The State in which risks associated with aviation activities, related to, or in direct support of the operation of aircraft, are reduced and controlled to an acceptable level.[190]

There is no reference to 'unlawful interference'. Rather, measures based on 'safety' should reduce risks to an acceptable level. The risks being those connected to the operation of the aircraft, rather than external factors based on intent like unlawful interference. The focus of each is then completely different and this is extremely important to appreciate when regulating cybersecurity as the implementation of the measure will be different depending on which regime the measure is found.

Upon observing air law provisions that utilise the term 'security', it becomes apparent that they are concerned with protection against man-made threats based on a level of intent; "security encompasses the threats of crime, terrorism and hostile human acts directly or indirectly affecting the physical integrity of the public and individuals".[191] This conforms to the definition provided in Regulation 300/2008 and its overall objective to protect aviation from unlawful interference by external unauthorised sources.

188 Ann. 17 to the Chicago Convention, 2017, at 1-2.

189 *See*, Chapter III.4.3.

190 Ann. 19 to the Chicago Convention, Safety Management, Second Edition, July 2016, at 1-2.

191 S. Kaiser, 'Space Security and Space Safety: Two Concepts to be Distinguished', 61st International Astronautical Congress, Prague, Czech Republic, IAC-10.E7.3.1 (2010), at 2.

Upon observing air law provisions that utilises the term 'safety', it becomes apparent that they concerned with protection against technical failures and human factors by authorised personnel.

> [S]afety relates to the design and quality of products and systems, and the applicable procedures, including human factors, for the use and maintenance of those products and systems. Technical safety is typically achieved by the establishment of technical rules, norms or standards and procedures for their compliance and verification.[192]

This conforms to the overall objective of Regulation 216/2008, which is safety based.

The distinction is, however, blurred and sometimes confused. This can be exemplified by Standard 2.1 in Annex 17 which specifies the objectives of the Annex.

> 2.1.1 Each Contracting State shall have as its primary objective the safety of passengers, crew, ground personnel and the general public in all matters related to safeguarding against acts of unlawful interference with civil aviation.
>
> 2.1.2 Each Contracting State shall establish an organization and develop and implement regulations, practices and procedures to safeguard civil aviation against acts of unlawful interference taking into account the safety, regularity and efficiency of flights.
>
> 2.1.3 Each Contracting State shall ensure that such an organization and such regulations, practices and procedures:
>
> a. protect the safety of passengers, crew, ground personnel and the general public in all matters related to safeguarding against acts of unlawful interference with civil aviation; and
>
> b. are capable of responding rapidly to meet any increased security threat.

Thus, the security-based annex has the primary objective of safety. It could then seem appropriate to classify 'security' as a subcategory of 'safety' whereby a level of intent is involved.

By observing the usage of the terms in air law, it becomes clear that they are used to denote different systems of rights and obligations. The division is not linguistic, but rather practical and has consequences. Therefore, when 'safety' is used in EU aviation law, it refers to the rights and obligations stemming from Regulation 216/2008 and Reg-

192 *Ibid.*, at 3.

ulation 2018/1139. When 'security' is used, it is referring to the rights and obligations stemming from Regulation 300/2008. In cases of overlap, such as where the same rights and obligations are found in both legal regimes, the conclusion must be that each source regulates independently and focuses within either 'safety' or 'security'.

9.3.4 Context: Regulation 2018/1139

In practice, Recitals 12 and 59 show that cybersecurity is to be dealt with by the EU aviation security regime as laid out by Regulation 300/2008,[193] and that the Agency via the EU aviation safety regime shall have a supportive role. Cybersecurity is security and security may impact safety.

While the Recitals have set the scene, the implementation of this by Articles 4 and 88, as well as generally in other Articles and the Annexes to Regulation 2018/1139, may present a different approach.

9.4 Principles for Measures Under This Regulation

The first text that could potentially directly regulate cybersecurity is found in Article 4(1)(c):

> 1. When taking measures under this Regulation the Commission, the Agency and the Member States shall
> [...]
> (c) allow for immediate reaction to established causes of accidents, serious incidents and intentional security breaches.

As noted above, the term 'security' can include cybersecurity. The inclusion of 'intentional' enforces the link to cybersecurity because there must be an attacker. In other words, there is a level of intent. The level of intent required must, therefore, be established.

The word 'cybersecurity' appears for the first time in the substantive Articles of the Regulation in Article 4(1)(d).

> take into account interdependencies between the different domains of aviation safety, and between aviation safety, cyber security and other technical domains of aviation regulation.

193 *See*, Chapter II.10.

Subsection (d) must be considered in two parts for cybersecurity.

First, it covers the interdependencies between the different domains of aviation safety. Thus, it will include aspects of cybersecurity, as a branch of security, where it has a direct link to safety.

Second, this scope is broadened as it covers the interdependencies between aviation safety and cybersecurity. This affirms the position taken in Recital 12 as it shows that the EU distinguishes aviation safety from cybersecurity. It is not clear, however, whether the EU is placing cybersecurity under the title 'security', as Recital 59 does, or if it considers it to constitute a domain of its own.

The obligation set out in Article 4 is, therefore, to *take into account* the relationship between different areas of safety, and the relationship between safety and cybersecurity. The obligation is not that the Agency should act upon the conclusions reached after taking such considerations. In other words, the Agency should merely appreciate the wider picture when taking its safety measures provided under the Regulation.

9.5 Interdependencies Between Civil Aviation Safety and Security

Article 88 was introduced into the EU aviation safety regime to address "technical aspects of aviation security directly linked with safety".[194]

9.5.1 Article 88(1)

Article 88(1) declares:

> The Commission, the Agency and the Member States shall cooperate on security matters related to civil aviation, including cyber security, where interdependencies between civil aviation safety and security exist.

This acknowledges the interdependencies that may be present between aviation security measures and aviation safety. The example provided in the regulatory proposal for Regulation 2018/1139 cites cockpit door locking systems, the relevance of which came to light following the 9/11 attacks in the US.[195] The doors are now designed to prevent unauthorised access to the cockpit, notably from hijackers,[196] with the requirements being established in Annex 6 to the Chicago Convention under Chapter 13 on 'Secur-

194 Proposal, COM/2015/0613 Final – 2015/0277 (COD), 7 December 2015, at 17.
195 *Ibid.*, at 17.
196 *See*, Chapter III.7.

ity'.[197] Hijacking and Chapter 13 are traditionally 'security' based, whereas Annex 6 is traditionally 'safety' based, thus succinctly demonstrating the interdependency.

The link between cybersecurity and security is again repeated, with the former being explicitly categorised as a subcategory of the latter. In the 2015 Proposal, this was the only time 'cybersecurity' featured in the text and was included as a political statement to make it explicitly clear that the Regulation includes 'cybersecurity' under the heading 'security'. As a result, Article 88(1) makes it explicitly clear under which domain cybersecurity is housed.

Article 88(1) obliges the Member States, the Commission and EASA to cooperate. The scope, context and procedure for cooperation are not specified, but the product of RMT.0720, an aviation cybersecurity Commission Regulation, will likely clarify these points.

9.5.2 Article 88(2)

Article 88(2) provides EASA with the competencies to provide technical assistance to the Commission in the implementation of Regulation 300/2008 when there is a link between safety and cybersecurity.

> Where interdependencies between civil aviation safety and security exist, the Agency shall, upon request, provide technical assistance to the Commission, where the Agency has the relevant safety expertise, in the implementation of Regulation (EC) No. 300/2008 of the European Parliament and of the Council and other relevant provisions of Union legislation.

This strengthens the bridge between traditional aviation security matters, as governed by Regulation 300/2008, and aviation safety, as governed by Regulation 2018/1139. As a result, the two legal texts must be compatible.

An example of the overlap between the two legal texts can be seen in Regulation 300/2008, Annex, Point 10.2 on In-Flight Security Measures.

---

197 Ann. 6 to the Chicago Convention, Operation of Aircraft, Vol. I – International Commercial Air Transport – Aeroplanes, Tenth Edition, July 2016, at 13.2.2:

> All passenger-carrying aeroplanes of a maximum certificated take-off mass in excess of 45,500 kg or with a passenger seating capacity greater than 60 shall be equipped with an approved flight crew compartment door that is designed to resist penetration by small arms fire and grenade shrapnel, and to resist forcible intrusions by unauthorized persons. This door shall be capable of being locked and unlocked from either pilot's station.

> Appropriate security measures such as training of flight crew and cabin staff shall be taken to prevent acts of unlawful interference during a flight.

The training of flight crew and cabin staff, as noted by Regulation 300/2008, links to Regulation 1178/2011, which lays down technical requirements and administrative procedures related to civil aviation aircrew pursuant to Regulation 216/2008 and Regulation 2018/1139.[198] It is possible that security training of flight crew and cabin staff can be incorporated into the mandatory training received per Regulation 1178/2011. This point may be taken up by RMT.0720 and subsequently added to EU secondary law.

9.5.3 Article 88(3)

Article 88(3) links measures traditionally taken by the Agency on aviation safety, to acts of unlawful inference.

> To contribute to protecting civil aviation against acts of unlawful interference where interdependencies between civil aviation safety and security exist, the Agency shall where necessary, react without undue delay to an urgent problem which is of common concern to Member States and which falls within the scope of this Regulation by:
>
> a. taking measures under point (h) of Article 77(1) to address vulnerabilities in aircraft design;
> b. recommending corrective actions to be taken by the national competent authorities or natural and legal persons subject to this Regulation and/or disseminating relevant information to those authorities and persons, in the case that the problem affects aircraft operation, including the risks to civil aviation arising from conflict zones.
>
> Before taking the measures referred to in points (a) and (b) of the first subparagraph, the Agency shall obtain the agreement of the Commission and consult the Member States. The Agency shall base those measures, where possible, on common Union risk assessments and take into account the need for rapid reaction in emergency cases.

198 Commission Regulation (EU) No. 1178/2011.

The Explanatory Memorandum states that this establishes a process to allow EASA "to take measures which are within the scope of its competence, such as airworthiness directives[199] or safety information bulletins,[200] to protect civil aviation against acts of unlawful interference".[201]

Corrective actions are dealt with under Article 76(6)(a).

> The Agency shall react without undue delay to an urgent safety problem falling within the scope of this Regulation by:
>
> a. determining corrective action to be taken by natural and legal persons in respect of which it acts as the competent authority and by disseminating related information to those persons, including directives or recommendations, where this is necessary to safeguard the objectives set out in Article 1; the Agency may also issue safety bulletins containing non-binding information or recommendations addressed to other natural and legal persons involved in aviation activities.

Article 76(6)(a) was included in the Regulation to allow the Agency to correct, with urgency, a safety problem falling within the scope of the Regulation. As the Agency should, pursuant to Article 88(3), "obtain the agreement of the Commission and consult the Member States" before taking corrective action, there is a contradiction between Article 76(6)(a) which requires prior agreement and Article 88(3) which allows for unilateral action by the Agency. The *lex specialis* nature of Article 76(6) means it would prevail, so agreement must be sought from the Commission.

Article 77(1) allows the Agency to "carry out on behalf of Member States the functions and tasks of the state of design, manufacture or registry, when those functions and tasks are related to design certification and mandatory continuing airworthiness information". Relevant for Article 88(3) is Article 77(1)(h), EASA shall:

---

199 "In accordance with Commission Regulation (EC) No. 1321/2014 (Annex I, M.A.301), the continuing airworthiness of an aircraft shall be ensured by accomplishing any applicable ADs. Consequently, no person may operate an aircraft to which an AD applies, except in accordance with the requirements of that AD unless otherwise specified by the Agency (Annex I, M.A.303) or agreed with the Authority of the State of Registry (Article 14(4) exemption of the Basic Regulation)." EASA, 'Airworthiness Directives', https://www.easa.europa.eu/easa-and-you/aircraft-products/airworthiness-directives-ad.

200 "An EASA Safety Information Bulletin (SIB) is an information tool that intends to alert, inform and draw the attention of the aviation community on safety issues. SIBs contain non-mandatory information and guidance that do not qualify for an Airworthiness Directive (AD)." EASA, Safety Information Bulletin (SIB), PR.CAP.00007-004, 1 July 2015, at 4.

201 Proposal, COM/2015/0613 Final – 2015/0277 (COD), 7 December 2015, at 17.

> ensure the continuing airworthiness functions associated with the design of products, the design of parts, non-installed equipment and equipment to control unmanned aircraft remotely it has certified and in respect of which it performs oversight, including reacting without undue delay to a safety or security problem and issuing and disseminating the applicable mandatory information.

Therefore, to ensure the continuing airworthiness of the subject matter in Article 77(1)(h), the Agency can react to safety and security issues, as well as share certain information. The security issue must have a link to safety, in that it is connected to the continuing airworthiness.

The relationship between Article 88(3) and the rest of the Regulation is unclear as it refers to 'unlawful interference'. This term is neither defined nor does it feature anywhere else in the Regulation. Article 88(3) is only applicable to actions taken to contribute to the protection of unlawful interference. Thus, it is narrower than 'safety' or 'security', and recourse is needed to Regulation 300/2008 and Annex 17 of the Chicago Convention as these utilise the term.[202] As a result, the scope of this subsection is unclear and its practical application currently unknown.

9.6 Essential Requirements to Regulation 2018/1139

Attached to the substantive Articles of Regulation 2018/1139 are the ten Annexes. The term 'cyber' does not feature in any of the Annexes, however, traces of cybersecurity can be found throughout.

First, the term 'security' can be found. Due to the link made in Article 88 between security and cybersecurity, it is possible to interpret the use of 'security' in the Annexes to include, where applicable, cybersecurity. Recital 20 of Regulation 2018/1139 states:

> In view of the increasing reliance of civil aviation on modern information and communication technologies essential requirements should be laid down to ensure the security of information used by the civil aviation sector.

The user of both 'security' and 'information' within the context of essential requirements clearly has relevance for cybersecurity.

202 *See*, Chapter II.10 and Chapter III.4.

Second, the use of 'unlawful interference' as found in Article 88 is missing from the Annexes.[203] However, five different formulations of interference are used.

1. 'Interference'.
2. 'Intentional and non-intentional system interference'.
3. 'Unauthorised interference'.
4. 'Harmful interference'.
5. 'Interference by electronic means'.

The use of 'interference' is the broadest of all the formulations. The other formulations have qualifiers that reduce their scope, notably whether there is intent, authorisation and if it causes harm. As 'unlawful interference' does not feature, there is no qualifier for the act to be illegal. The different qualifiers, thus mean different things and should be treated accordingly.

Third, the Annex also introduces additional terms that can relate to cybersecurity. For example, Annex II 1 on the Essential Requirements for Airworthiness states that "[p]roduct integrity, including protection against information security threats, must be assured for all anticipated flight conditions for the operational life of the aircraft". No definition or clarifications are offered to the scope and meaning of the term 'information security threats', however, this is a common term in cybersecurity to denote attack methods. The term 'information security threats' appears only twice in the Essential Requirements under Annex II 1 and 1.3.5. This phrase was included, in order to make the provision as broad as possible. This is because 'cybersecurity threats' would be limited to electronic threats, whereas, 'information' would include both electronic and physical security threats.

Fourth, in addition to phraseology, there are words used throughout that can have cybersecurity connotations depending upon the context. For example, Annex VIII on the Essential Requirements for ATM/ANS and Air Traffic Controllers hosts the following terms in 2.4:

1. Availability.
2. Integrity.
3. Continuity.
4. Timeliness.
5. Corruption.

203 *See*, Regulation 2018/1139, Art. 74(8)(f).

The same Annex under 2.2.3 also uses the word 'reliable' within the scope of means of communication. These words are prevalent in cybersecurity and have been included in the Regulation, *inter alia*, for this reason.[204]

Fifth, standard aviation safety clauses can also be read to include cybersecurity. For example, Annex II 1.3.5 states that "[d]esign precautions must be taken to minimise the hazards to the aircraft and occupants from reasonably probable threats". The term 'reasonably probable threat' is not defined in the Regulation. However, due to the extensive steps taken by the EU to regulate aviation cybersecurity, with explicit links to safety being mentioned in Regulation 2018/1139, and due to the increasing number of attacks taking place against aviation stakeholders as outlined in Table 1, a cyberattack may be defined as a reasonably probable threat. Therefore, design precautions should take cybersecurity into consideration.

Sixth, under Annex II 1.3.2, the aircraft, its systems and equipment 'must function as intended'. This can be seen as an implicit requirement, so precautions do not have to be designed-in against non-intended functionality or as an explicit requirement, thus requiring additional steps to provide the intended functionality. The interpretation of 'intended' can also produce different readings of the requirement. 'Intended' could be read objectively to mean in the scope of Regulation 2018/1139. This would then place a cyberattack that has a link to safety within the scope of 'as intended'. Alternatively, 'intended' could be read subjectively to mean that which is intended by the operator, owner or pilot. If the aircraft, systems and equipment are functioning as the relevant person requires, but a cyberattack is happening, such as a passive cyberattack, then this would fall outside the scope of 'as intended'. Cybersecurity can, thus, be read into general provisions of the Annexes.

Although there are numerous potential links between the Essential Requirements and cybersecurity, they must be read alongside the Recitals, of which Recitals 12 and 59 make it explicitly clear that cybersecurity is a branch of security and not safety. Therefore, the Essential Requirements, which are safety based, would appear to be separate from cybersecurity unless a distinction can be made between cybersecurity and the terms found in the Annexes (e.g. 'cybersafety').

The same line of argumentation can be applied to Regulation 216/2008 and its Annexes as it also contains terms and wording that can be interpreted to include elements of cybersecurity. For example, Regulation 216/2008 uses the term 'interference' six times and Annex Va on the Essential Requirements for Aerodromes contains the words 'rele-

204 *See*, for way of a comparison, its link with Art. 34 of the Chicago Convention.

vant', 'accurate', 'readable', 'complete', 'unambiguous', 'available' and 'secure' within the context of data, so these could have cybersecurity relevance.

9.7 *Implementing Rules*

The same line of argumentation as found in Chapter II 9.6 can be extended to the Implementing Rules, which initially had Regulation 216/2008 as their legal basis, that fall into the following headings.

1. Initial airworthiness.
2. Continuing and additional airworthiness.
3. Air crew.
4. Air operations (including balloons, sailplanes).
5. Third country operators.
6. Air traffic management and air navigation services.
7. Air traffic controllers.
8. Airspace usage requirements.
9. Standardised European Rules of the Air (SERA).
10. Aerodromes.
11. Other (administrative).[205]

For example, Regulation 2017/373, as noted in Chapter II.2.2.4, makes a specific reference to the term 'cybersecurity'. Further, Annex I 4(a) of Regulation 1035/2011 on the common requirements for the provision of ANS[206] states that ANS "providers shall establish a security management system to ensure: (a) the security of their facilities and personnel so as to prevent unlawful interference with the provision of air navigation services".[207] The term 'unlawful interference' could provide the link to cybersecurity. However, the categorisation in the Recitals of cybersecurity and a branch of security should, again, not be overlooked.

---

205 *See*, for the full list, EASA, 'Regulations', https://www.easa.europa.eu/regulations. *See*, P. Mendes de Leon, 2017, Ch. 7.

206 Commission Implementing Regulation (EU) No. 1035/2011 of 17 October 2011 Laying Down Common Requirements for the Provision of Air Navigation Services and Amending Regulations (EC) No. 482/2008 and (EU) No. 691/2010, OJ L 271, 18 October 2011, pp. 23-41.

207 *See*, for further examples of 'unlawful interference', Commission Implementing Regulation (EU) No. 923/2012 of 26 September 2012 Laying Down the Common Rules of the Air and Operational Provisions Regarding Services and Procedures in Air Navigation and Amending Implementing Regulation (EU) No. 1035/2011 and Regulations (EC) No. 1265/2007, (EC) No. 1794/2006, (EC) No. 730/2006, (EC) No. 1033/2006 and (EU) No. 255/2010, OJ L 281, 13 October 2012, pp. 1-66, at SERA.11001; and Commission Regulation (EU) No. 677/2011 of 7 July 2011 Laying Down Detailed Rules for the Implementation of Air Traffic Management (ATM) Network Functions and Amending Regulation (EU) No. 691/2010, OJ L 185, 15 July 2011, pp. 1-29, at Ann. V 3.

The role of EU Regulation within the EU is not only relevant for EASA. For example, Eurocontrol is an international organisation consisting of 41 European Member States (as per July 2019), which provides assistance to EASA with the drafting safety-based implementing rules. It works with its Member States to help air traffic operations throughout Europe. The organisations "activities involve all gate-to-gate air navigation service operations: strategic and tactical flow management, controller training, regional control of airspace, safety-proofed technologies and procedures, and collection of air navigation charges".[208] This, therefore, includes the EU Regulations on air traffic management and air navigation services.

NPA 2019-01 marks the first change to the Implementing Rules system (regulations, AMC, GM and CS). Therefore, NPA 2019-01's wording must be assessed to see how EASA is addressing cybersecurity in practice. CS 25.1319(a) on Equipment, Systems and Network Information Security Protection will be used as an example.

> a. Aeroplane equipment, systems and networks, considered separately and in relation to other systems, must be protected from intentional unauthorised electronic interactions that may result in adverse effects on the safety of the aeroplane. Protection must be ensured by showing that the security risks have been identified, assessed and mitigated as necessary.

It can be observed in the first sentence that EASA avoids referring to cybersecurity, safety or security when specifying the subject matter of the CS, rather it refers to "intentional unauthorised electronic interactions". Therefore, the act can be:

1. either lawful or unlawful, so long as it is intentional,
2. performed by an external entity or from an insider, so long as it is not authorised,
3. any attach method, so long as it is electronic,
4. any attach method, so long as there is an interaction.

While the scope of the CS is clearly limited by the wording, such as passive attack methods may not be included, EASA avoids the discussion on safety/security. However, this is not the case for the second sentence where EASA is clearly covering security aspects. This is a good example of EASA's interpretation of its mandates; safety, as well as security aspects that impact on safety.

The products of RMT.0720 will also impact the Implementing Rules as the existing ones may be amended or new Regulations may be produced.

208 R. Abeyratne, *Air Navigation Law*, First Edition, (Springer, 2012), at 225.

10 Aviation Security in the European Union

10.1 Introduction to the Aviation Safety Regulation

The development of regional security-based measures stems back to ECAC Document 30,[209] which in 1985 established the first uniform standards for security for air transport in Europe. Security is a core activity of ECAC with its work centred around "ensuring security measures are in place to protect civil aviation against acts of unlawful interference".[210] One method is to describe aviation security measures that are contained in ECAC Document 30 Part II, which include:

1. ground-based provisions, such as securing airport supplies, baggage and mail, and
2. in flight provision, including ATM cybersecurity and the management of hijackings.

The content of Document 30 Part II reflects that of Annex 17;[211] however, due to its ongoing revisions and updates, it is sometimes more extensive. As ECAC has no legislative powers, Document 30 has no binding legal effect and, therefore, lacks a legal enforcement mechanism.[212]

Prior to the terrorist attacks that took place in the US on 11 September 2001, aviation security within the EU was managed at either a national level or by ICAO through Annex 17 with the support of Document 30 Part II. However, following the attacks, aviation security was placed high on the EU agenda. Consequently, Regulation 2320/2002, which established the common rules in the field of civil aviation security was adopted, as a framework Regulation, "to establish and implement appropriate Community measures, in order to prevent acts of unlawful interference against civil aviation".[213] The content of this Regulation was based on Document 30 Part II. Regulation 2320/2002 marked a significant step in the EU's efforts to regulate aviation security as it provided the foundations for an extensive and detailed set of rules covering 'unlawful interference'.

The wording of the Regulation, both in its title and in its primary objective set out in Article 1(1), showed a strong connection to the Chicago Convention security regime and Document 30 Part II, as the focus of the Regulation was on 'unlawful interference'. This

---

209 ECAC Policy Statement in the Field of Civil Aviation Facilitation, ECAC Doc. No. 30 (Part I), Eleventh Edition, December 2009. *See*, for more information on ECAC, Ch. III.XX.

210 ECAC, 'Security', https://www.ecac-ceac.org/security.

211 *See*, Chapter III.4.2.

212 J. Faust and C. Leininger, 'Security', *in* S. Hobe, N. von Ruckteschell and D. Heffernan (eds), *Cologne Compendium on Air Law in Europe*, English Edition, (Carl Heymanns Verlag, 2014), at 782.

213 Regulation (EC) No. 2320/2002 of the European Parliament and of the Council of 16 December 2002 Establishing Common Rules in the Field of Civil Aviation Security, OJ L 355, 30 December 2002, pp. 1-23.

connection is strengthened as the secondary objective, as set out in Article 1(2), "is to provide a basis for a common interpretation of the related provisions of the Chicago Convention, in particular its Annex 17". Therefore, the wording of the Regulation is based on Annex 17 and it is designed, as an interpretive tool, to form part of the international security regime.

Regulation 2320/2002 set the foundations for aviation security regulation in Europe and Chapter II.10.3 will focus on its regulatory legacy and see the relevance to cybersecurity. Regulation 2320/2002 was not designed to be static, rather it was designed to evolve with the changing threat landscape.

> Given the need for more flexibility in adopting security measures and procedures in order to meet evolving risk assessments and to allow new technologies to be introduced, this Regulation should lay down the basic principles of what has to be done in order to safeguard civil aviation against acts of unlawful interference without going into the technical and procedural details of how they are to be implemented.[214]

Therefore, the rules should be "regularly updated to address evolving risks and threats as well as technological changes".[215]

> The content of Regulation (EC) No. 2320/2002 should be revised in the light of the experience gained, and the Regulation itself should be repealed and replaced by this Regulation seeking the simplification, harmonisation and clarification of the existing rules and the improvement of the levels of security.[216]

In March 2008, it was repealed and replaced by Regulation 300/2008 on the common rules in the field of civil aviation security, which came into full effect from 29 April 2010.

10.2 European Civil Aviation Conference

Due to the relevance of ECAC for the origins of EU aviation security, it will be looked at in more detail here. ECAC is an intergovernmental organisation that was founded in 1955. Its aim is to "harmonise civil aviation policies and practices amongst its Member States and, at the same time, promote understanding on policy matters between its Mem-

214 Regulation 300/2008, Recital 5.
215 EU Parliament, 'Air Transport: Civil Aviation Security', Fact Sheets on the European Union', February 2018, http://www.europarl.europa.eu/atyourservice/en/displayFtu.html?ftuId=FTU_3.4.7.html.
216 Regulation 300/2008, Recital 4.

ber States and other parts of the world".[217] ECAC's membership consists of 44 European Member States (as of July 2019).[218] ECAC does not have law-making powers, however, its role in information building and the dissemination of information is important. ECAC has taken four key steps with regard to aviation cybersecurity.

First, ECAC Document 30 Part II sets security policy and these "recommendations are made to [the] Director General [of] Civil Aviation for inclusion in State security programmes and are promulgated by ECAC Document 30".[219] Chapter 14 on Cyber Security was introduced in 2010 to specifically address cybersecurity. This is not publicly available so further comments cannot be made.

Second, ECAC is engaged in international outreach where it shares information on cybersecurity. This has included sharing information with ICAO, assisting with provisions on cybersecurity in ICAO Annex 17, and aiding with a working paper during the 39th ICAO Assembly.[220]

Third, since 2009 ECAC has had a Study Group on Cyber Threats to Civil Aviation, which consists of the ECAC Member States and has observers that include ACI, Airbus, EASA, EU, Eurocontrol, IATA, IFALPA and the US.[221]

Fourth, ECAC has a Capacity Building Programme, which has conducted vulnerability assessments on cybersecurity since 2011 and has held multilateral workshops with the aim of providing awareness training.[222]

ECAC is, therefore, active in researching cybersecurity issues and spreading awareness of the issue.

10.3 *Objectives of Regulation 300/2008*

The first objective of the Regulation, as set out in Article 1(1), is to establish "common rules to protect civil aviation against acts of unlawful interference that jeopardise the security of civil aviation". This can then be divided into three focal sections which will be analysed below in Chapter II.10.3.1 to Chapter II.10.3.3. The second objective is also

217 ECAC, 'About ECAC', https://www.ecac-ceac.org/about-ecac.

218 ECAC, 'Member States', https://www.ecac-ceac.org/member-states.

219 European Regions Airline Association, 'ECAC', https://www.eraa.org/policy/security/ecac.

220 P. Reverdy, Deputy Executive Secretary ECAC, 'ECAC Priorities for Cyber Security: Building on our Strengths', ICAO Cyber Summit and Exhibition, Dubai, UAE, 5-6 April 2017.

221 *Ibid.*

222 *Ibid.*

provided in Article 1(1) whereby the Regulation "provides the basis for a common interpretation of Annex 17". Therefore, any cybersecurity elements in Annex 17 have a link to Regulation 300/2008. In order to achieve these two objectives, the EU shall pursuant to Article 1(2), set "common rules and common basic standards on aviation security" and "mechanisms for monitoring compliance".

10.3.1 Protect Against

Regulation 300/2008 sets to 'protect against' and this relates to, pursuant to Article 1(2)(a) and (b), the "setting of common rules and common basic standards on aviation security", and "mechanisms for monitoring compliance". Therefore, protection is envisaged through legal controls and ensuring that they are implemented.

10.3.2 Unlawful Interference

The Regulation sets to protect against 'unlawful interference'. While the Regulation does not define 'unlawful interference', the definition provided by ICAO in Annex 17 in Chapter 1 can be used as an interpretative tool.[223]

Annex 17 provides a definition of the term 'unlawful interference' in Chapter 1.

> Acts of unlawful interference. These are acts or attempted acts such as to jeopardize the safety of civil aviation, including but not limited to:
> unlawful seizure of aircraft,
> destruction of an aircraft in service,
> hostage-taking on board aircraft or on aerodromes,
> forcible intrusion on board an aircraft, at an airport or on the premises of an aeronautical facility,
> introduction on board an aircraft or at an airport of a weapon or hazardous device or material intended for criminal purposes,
> use of an aircraft in service for the purpose of causing death, serious bodily injury, or serious damage to property or the environment,
> communication of false information such as to jeopardize the safety of an aircraft in flight or on the ground, of passengers, crew, ground personnel or the general public, at an airport or on the premises of a civil aviation facility.

The use of Annex 17 as an interpretive tool is justifiable as the Regulation, pursuant to Article 1(1), is to provide a "basis for a common interpretation of Annex 17". Therefore, there is a link between the two documents in which interpretation is a central focus and also such argumentation is furthered as the Regulation was drafted within the context of

223 *See*, Chapter III.4.3.

the Chicago Convention regime, to build upon it, rather than for it to exist in isolation. As a result, the Regulation is to be seen as being compliant with Annex 17 and a tool to supplement it by prescribing EU-wide standards for aviation security.

The act needs to be unlawful, so it must be prohibited by a source of law. Rectal 7 takes note of other security-based rules.

> Without prejudice to the Convention on offences and certain other acts committed on board aircraft, Tokyo, 1963, the Convention for the suppression of unlawful seizure of aircraft, The Hague, 1970, and the Convention for the suppression of unlawful acts against the safety of civil aviation, Montreal, 1971, this Regulation should also cover security measures that apply on board an aircraft, or during a flight, of Community air carriers.

Thus, the intention of the Regulation is for it to fall within the existing international legal body of laws for aviation security. Therefore, compiling all of this, 'unlawful interference' includes, *inter alia*, hijacking and sabotage when prohibited by law. The scope of 'unlawful interference' within the Regulation may, however, be broader than the ICAO regime as other sources of law, including national law, are not excluded. There is, however, the qualifier that the act must be unlawful. This can limit the scope of the regulation within the context of cyberattacks as "not all cybersecurity issues affecting the safety of civil aviation are unlawful and/or intentional, and should, therefore, be addressed through the application of safety management systems".[224]

The act must also cause interference. This term is undefined and harder to interpret within the context of cybersecurity. A cyberattack that affects the victim physically, such as controlling the aircraft, can be more easily interpreted as interference. This is supported by the fact that acts, such as hijacking and sabotage are included within the definition of 'unlawful interference'. However, it is less clear whether eavesdropping, tapping, key logging and other passive attack methods constitute interference.[225] Passive attacks may interact with the technology or a resulting product, such as communication links, but do not physically interfere with the victim and the victim's use of that technology. As a result, it is not certain whether such acts constitute interference under Regulation 300/2008. Therefore, it is argued that this term has serious weaknesses and, because of this, it is not suitable for regulating cybersecurity.

224 ICAO, Assembly – 39th Session, Executive Committee, Addressing Cybersecurity in Civil Aviation, A39-WP 17 EX/5, 30 May 2016.

225 *See*, Chapter I.2.2.1.

10.3.3 Must Jeopardise the Security of Civil Aviation

The unlawful interference 'must jeopardise the security of civil aviation'. This limits the scope of the Regulation to just those acts of unlawful interference with a security implication. Article 3(2) defines aviation security.

> '[A]viation security' means the combination of measures and human and material resources intended to safeguard civil aviation against acts of unlawful interference that jeopardise the security of civil aviation.

The unlawful interference must, in order to fall within a strict interpretation of Article 1 (1), jeopardise the measures and resources intended to stop the unlawful interference, as this would *stricto sensu* be all that is incorporated within the scope of 'security'. This would drastically limit the scope of the Regulation and produce a large lacuna as acts of unlawful interference against aviation stakeholders would not be covered. While this may appear to be a shortcoming, the Regulation is applied in a way that circumvents this issue as it prescribes stakeholder targeted measures.

10.4 Scope of Regulation 300/2008

The Regulation establishes, in Article 2(1), which stakeholders the targeted measures are applicable.

> 1. This Regulation shall apply to the following:
>   a. all airports or parts of airports located in the territory of a Member State that are not exclusively used for military purposes;
>   b. all operators, including air carriers, providing services at airports referred to in point (a);
>   c. all entities applying aviation security standards that operate from premises located inside or outside airport premises and provide goods and/or services to or through airports referred to in point (a).

It is evident that this is limited to those stakeholders that have a physical link to the aircraft – both EU and non-EU operators – before and after it is engaged in flight involving an airport in the territory of a Member State. Therefore, the Regulation does not cover all aviation stakeholders as identified in Chapter I.3. Despite these limitations, the Regulation may still have relevance for combating cyberattacks against certain aviation stakeholders.

10.5 Common Basic Standards

Article 1(2) provides two mechanisms aimed towards protecting civil aviation against unlawful interference.

> The means of achieving the objectives set out in paragraph 1 shall be:
> a. the setting of common rules and common basic standards on aviation security;
> b. mechanisms for monitoring compliance.

These mechanisms will be at the centre of this subchapter and will be further investigated below to establish their relevance to aviation cybersecurity.

Pursuant to Article 4(1) "[t]he common basic standards for safeguarding civil aviation against acts of unlawful interference that jeopardise the security of civil aviation shall be as laid down in Annex I". The content of Annex I then comprises of measures designed to protect aviation from unlawful interference. Article 4 divides these common basic standards into two categories. First, Article 4(2) covers general measures.

> a. methods of screening allowed;
> b. categories of articles that may be prohibited;
> c. as regards access control, grounds for granting access to airside and security restricted areas;
> d. methods allowed for the examination of vehicles, aircraft security checks and aircraft security searches;
> e. criteria for recognising the equivalence of security standards of third countries;
> f. conditions under which cargo and mail shall be screened or subjected to other security controls, as well as the process for the approval or designation of regulated agents, known consignors and account consignors;
> g. conditions under which air carrier mail and air carrier materials shall be screened or subjected to other security controls;
> h. conditions under which in-flight supplies and airport supplies shall be screened or subjected to other security controls, as well as the process for the approval or designation of regulated suppliers and known suppliers;
> i. criteria for defining critical parts of security restricted areas;
> j. criteria for staff recruitment and methods of training;
> k. conditions under which special security procedures or exemptions from security controls may be applied; and

l. any general measures designed to amend non-essential elements of the common basic standards referred to in paragraph 1 by supplementing them not foreseen at the date of entry into force of this Regulation.[226]

Second, Article 4(3) covers detailed measures.

a. requirements and procedures for screening;
b. a list of prohibited articles;
c. requirements and procedures for access control;
d. requirements and procedures for the examination of vehicles, aircraft security checks and aircraft security searches;
e. decisions to recognise the equivalence of security standards applied in a third country;
f. as regards cargo and mail, procedures for the approval or designation of, and the obligations to be fulfilled by, regulated agents, known consignors and account consignors;
g. requirements and procedures for security controls of air carrier mail and air carrier materials;
h. as regards in-flight supplies and airport supplies, procedures for the approval or designation of, and the obligations to be fulfilled by, regulated suppliers and known suppliers;
i. definition of critical parts of security restricted areas;
j. staff recruitment and training requirements;
k. special security procedures or exemptions from security controls;
l. technical specifications and procedures for approval and use of security equipment; and
m. requirements and procedures concerning potentially disruptive passengers.[227]

From a simple reading of the two lists, it is evident that no reference is made to cybersecurity measures. However, this does not mean that the measures are not relevant for cybersecurity.

The two lists can be grouped into four overarching titles, which will provide the structure of the analysis.

---

226 Regulation 300/2008, Art. 4(2).
227 *Ibid.*, Art. 4(3).

10.5.1 Restricted Areas

The first overarching measure is the creation of restricted areas and control over entry to specified areas. For example, Annex I Part 1 and 2 on Airport Security contains measures based on:

1. airport planning requirements,
2. access control,
3. screening of persons other than passengers and items carried,
4. examination of vehicles,
5. surveillance, patrols and other physical controls, and
6. restricted areas.

These are then supplemented by Commission Implementing Regulation 2015/1998.[228] Physical controls are a necessary part of cybersecurity, so this is relevant. However, these measures must be implemented with cybersecurity in mind, for example, to ensure authorised personnel are vigilant to people using unauthorised technology and that the physical controls actually defend against this type of attack. As no reference is made to cybersecurity, it is then for those responsible for the implementation of the Regulation to include this within the scope of the measure or for Member States to apply more stringent measures where necessary.

10.5.2 Prohibited Articles

Several of the measures involve the banning of 'prohibited articles' in security restricted areas. Article 2(7) defines 'prohibited articles' as "weapons, explosives or other dangerous devices, articles or substances that may be used to commit an act of unlawful interference that jeopardise the security of civil aviation". Therefore, the question must be asked whether an attack method or technology to implement a cyberattack constitutes a prohibited article per this definition.

Article 2(7) provides an insight into what amounts to 'prohibited articles' under the Regulation, however, within the context of cybersecurity, further investigation is required. Commission Regulation 272/2009, which supplements the common basic standards on civil aviation security laid down in the Annex I to Regulation 300/2008,[229]

---

228 Commission Implementing Regulation 2015/1998 then lays down the detailed measures for the implementation of the common basic standards on aviation security which must be consulted in parallel with Ann. I of Regulation 300/2008. Commission Implementing Regulation (EU) No. 2015/1998 of 5 November 2015 Laying Down Detailed Measures for the Implementation of the Common Basic Standards on Aviation Security, OJ L 299, 14 November 2015, pp. 1-142.

229 Commission Regulation (EC) No. 272/2009 of 2 April 2009 Supplementing the Common Basic Standards on Civil Aviation Security Laid Down in the Annex to Regulation (EC) No. 300/2008 of the European Parliament and of the Council, OJ L 91, 3 April 2009, pp. 7-13.

provides some clarity on this interpretive issue within Annex I Part B on the category of articles that may be prohibited.

> The implementing rules to be adopted pursuant to Article 4(3) of Regulation (EC) No. 300/2008 may prohibit, under defined conditions, the introduction of any or all of the following categories of articles into security restricted areas and on board an aircraft:
> a. guns, firearms and other devices that discharge projectiles – devices capable, or appearing capable, of being used to cause serious injury by discharging a projectile;
> b. stunning devices – devices designed specifically to stun or immobilise;
> c. objects with a sharp point or sharp edge – objects with a sharp point or sharp edge capable of being used to cause serious injury;
> d. workmen's tools – tools capable of being used either to cause serious injury or to threaten the safety of aircraft;
> e. blunt instruments – objects capable of being used to cause serious injury when used to hit; and
> f. explosives and incendiary substances and devices – explosives and incendiary substances and devices capable, or appearing capable, of being used to cause serious injury or to pose a threat to the safety of aircraft.

Commission Regulation 272/2009 has since been amended to include 'liquids, aerosols and gels'.[230] Nothing in this list can be interpreted to include cyberattack methods or related technologies, thus it can be concluded that this part of the EU aviation security regulatory system does not yet apply to cyberattacks.

There is a clear link here with Regulation 2018/1139 as, for example, Annex V 1.4 states:

> Articles or substances, which are capable of posing a significant risk to health, safety, property or the environment, such as dangerous goods, weapons and ammunition, must not be carried on any aircraft, unless specific safety procedures and instructions are applied to mitigate the related risks.

230 *See*, Commission Regulation (EU) No. 297/2010 of 9 April 2010 Amending Regulation (EC) No. 272/2009 Supplementing the Common Basic Standards on Civil Aviation Security, OJ L 90, 10 April 2010, pp. 1-3; Commission Regulation (EU) No. 720/2011 of 22 July 2011 Amending Regulation (EC) No. 272/2009 Supplementing the Common Basic Standards on Civil Aviation Security as Regards the Phasing-In of the Screening of Liquids, Aerosols and Gels at EU Airports, OJ L 193, 23 July 2011, pp. 19-21; and Commission Regulation (EU) No. 245/2013 of 19 March 2013 Amending Regulation (EC) No. 272/2009 as Regards the Screening of Liquids, Aerosols and Gels at EU Airports, OJ L 77, 20 March 2013, pp. 5-7.

As a result, due to the restrictions placed on the term 'prohibited articles', Annex V 1.4 does not apply to cybersecurity.

The position is less clear with regard to other terms. For example, as discussed in Chapter III.3, a cyberattack may constitute a 'weapon' under certain conditions, for example, if it causes the total loss of the aircraft in flight and where a technology used to implement a cyberattack may be defined as a 'dangerous device'. However, while a cyberattack may cause destruction, such as the Stuxnet attack,[231] this does not render the cyberattack as an 'explosive' as it is not a substance that can be made to explode like Dynamite. Without definitions or clarifications from the Court of Justice of the European Union, the scope of the terms is unclear within the context of cybersecurity.

10.5.3 Training

There is a need to train staff in cybersecurity as appropriate 'cyber hygiene' can dramatically reduce the risk of a cyberattack taking place. Therefore, Article 4(2)(j) and Article 4(3)(j) on the criteria and requirements of staff recruitment and training have significant relevance for ensuring cybersecurity.

Pursuant to Annex 11.1 of Regulation 300/2008:

> Persons implementing, or responsible for implementing, screening, access control or other security controls shall be recruited, trained and, where appropriate, certified so as to ensure that they are suitable for employment and competent to undertake the duties to which they are assigned.

Therefore, persons responsible for such functions should be screened during the recruitment stage and then trained to ensure that they can carry out their security-based work.

Annex 11.2 is linked with the demarcation of restricted areas as it requires that persons, other than passengers, who require access to security restricted areas receive training so that these persons can implement the correct security measures.

Annex 11.3 states that training should be initial and recurrent. This is an appropriate methodology as it ensures that persons are trained when starting the position of employment and that their training is continuously refreshed so they are updated with new practices. The initial training should then include relevant cybersecurity training. The recurrent training should be as frequent as required, thus refreshing the individual's cybersecurity knowledge and updating them on any evolving threats.

231 *See*, Chapter II.4.3.

Finally, Annex I 11.4 requires that instructors responsible for such training, as listed in paragraphs Annex 1 11.1 and 11.2, have the necessary qualifications. This is an appropriate measure to check whether the instructors have undergone the necessary training to provide students with the accepted level of skills to appropriately carry out their functions. Thus, the system ensures that the necessary persons are trained by qualified instructors. However, a weakness presents itself if the qualifications received by the instructor do not incorporate cybersecurity.

Annex I.11 of Commission Implementing Regulation 2015/1998 then provides detailed measures for the implementation of the common basic standards of which cybersecurity-based training can be worked into the actual training where necessary. There is again a link with Regulation 2018/1139, as its Annex IV also covers training. Pilot training is contained in Section 1 and cabin crew training is contained in Section 4.

10.5.4 Third Countries

The final main category of measurement is found in Article 20 of the Regulation and relates to agreements between the EU and third countries.

> When appropriate, and in conformity with Community law, agreements recognising that the security standards applied in a third country are equivalent to Community standards could be envisaged in aviation agreements between the Community and a third country in accordance with Article 300 of the Treaty, in order to advance the goal of 'one-stop security' for all flights between the European Union and third countries.

Regulation 2015/2426, which amends Regulation 2015/1998, covers third countries recognised as applying security standards equivalent to the common basic standards on civil aviation security. The following third countries:

1. Canada,
2. Faroe Islands, in regard to Vagar airport,
3. Greenland, in regard to Kangerlussuaq airport,
4. Guernsey,
5. Isle of Man,
6. Jersey,
7. Montenegro, and
8. US

have been recognised as applying security standards equivalent to the common basic standards on civil aviation security. Aviation cybersecurity is a global responsibility, where cooperation and high global standards are key, so this is a step in the right direction. While this is a positive step, due to the limited number of third countries and the limited scope of the Regulation to cybersecurity, this has a very limited relevance to cybersecurity.

10.6 Additional Common Basic Standards

It must be considered that Article 4(1) allows for the adoption of "[a]dditional common basic standards not foreseen upon the entry into force of this Regulation, which should be added to the Annex in accordance with the procedure referred to in Article 251 of the Treaty".[232] Therefore, the supporting Regulations must also be examined in case relevant measures on cybersecurity are found. One tool to facilitate this is the EU's Stakeholder Advisory Group for Aviation Security (SAGAS), membership of which consists of the European Commission (DG-MOVE), Member State security representatives and industry representatives. The role of SAGAS is to "advise the Commission in the preparation of legislative proposals and policy initiatives as well as in relation to the implementation of existing legislation".[233] Furthermore, there is the Regulatory Committee for Civil Aviation Security AVSEC, which was created by Article 19 of Regulation 300/2008 to address the evolving threat to civil aviation. As a result, additional common basic standards may be published in the future to promote cybersecurity.

10.7 More Stringent Measures and Derogations

Pursuant to Article 6(1), a "Member State may apply more stringent measures than the common basic standards referred to in Article 4". More stringent measures must be based on a risk assessment and compliant with EU law, as well as being "relevant, objective, non-discriminatory and proportional to the risk that is being addressed". Therefore, Member States are able to develop more stringent measures outside of Annex I, which could provide Member States with the basis to build on measures aimed towards tackling cybersecurity issues.

Another form of flexibility is provided by Commission Regulation 1254/2009,[234] which sets the criteria to allow Member States to derogate from the common basic standards on

232 The reference to 'the Treaty' corresponds to the TFEU.

233 Commission, 'Group Details – Commission Expert Group', http://www.ec.europa.eu/transparency/regexpert/index.cfm?do=groupDetail.groupDetail&groupID=2883.

234 Commission Regulation (EU) No. 1254/2009 of 18 December 2009 Setting Criteria to Allow Member States to Derogate from the Common Basic Standards on Civil Aviation Security and to Adopt Alternative Security Measures, OJ L 338, 19 December 2009, pp. 17-17.

civil aviation security and to adopt alternative security measures. Commission Regulation 2016/2096,[235] which amends Commission Regulation 1254/2009, states in Article 1:

> Member States may derogate from the common basic standards referred to in Article 4(1) of Regulation (EC) No. 300/2008 and adopt alternative security measures that provide an adequate level of protection on the basis of a risk assessment approved by the appropriate authority at airports or demarcated areas of airports where traffic is limited to one or more of the following categories.

Article 1 then lists a number of categories, based on the type of operation or type of aircraft, where this applies.[236] Member States may then integrate cybersecurity measures via this mechanism.

The EU aviation security regime then provides flexibility, while maintaining the approved EU standard, which could allow Member States to strengthen cybersecurity rules.

---

235 Commission Regulation (EU) No. 2016/2096 of 30 November 2016 Amending Regulation (EU) No. 1254/2009 as Regards Certain Criteria to Allow Member States to Derogate from The Common Basic Standards on Civil Aviation Security and to Adopt Alternative Security Measures, OJ L 326, 1 December 2016, pp. 7-8.

236 This is done by Regulation 1254/2009 which was amended by Regulation 2016/2096. The following categories include:

1. aircraft with a maximum take-off weight of less than 15 000 kilograms;
2. helicopters;
3. State, military and law enforcement flights;
4. fire suppression flights;
5. flights for medical services, emergency or rescue services;
6. research and development flights;
7. flights for aerial work;
8. humanitarian aid flights;
9. flights operated by air carriers, aircraft manufacturers or maintenance companies, transporting neither passengers and baggage, nor cargo and mail;
10. flights with aircraft with a maximum take-off weight of less than 45 500 kilograms, owned by a company for the carriage of own staff and non-fare-paying passengers and goods as an aid to the conduct of company business;
11. flights with aircraft with a maximum take-off weight of less than 45 500 kilograms, chartered or leased in its entirety by a company from an aircraft operator with which it has a written agreement for the carriage of own staff and non-fare-paying passengers and goods as an aid to the conduct of company business;
12. flights with aircraft with a maximum take-off weight of less than 45 500 kilograms, for the carriage of the owner of the aircraft and of non-fare-paying passengers and goods.

10.8 Implementation and Oversight

An important part of ensuring cybersecurity measures are adhered to is to provide adequate implementation measures. The Regulation has facilitated this via a series of targets programmes.

1. The national civil aviation security programme: "That programme shall define responsibilities for the implementation of the common basic standards referred to in Article 4 and shall describe the measures required by operators and entities for this purpose."[237]
2. The national quality control programme: "That programme shall enable the Member State to check the quality of civil aviation security in order to monitor compliance both with this Regulation and with its national civil aviation security programme."[238]
3. The airport security programme: "That programme shall describe the methods and procedures which are to be followed by the airport operator in order to comply both with this Regulation and with the national civil aviation security programme of the Member State in which the airport is located."[239]
4. The air carrier security programme: "That programme shall describe the methods and procedures which are to be followed by the air carrier in order to comply both with this Regulation and with the national civil aviation security programme of the Member State from which it provides services."[240]
5. The entity security programme: "That programme shall describe the methods and procedures which are to be followed by the entity in order to comply with the national civil aviation security programme of the Member State in respect of its operations in that Member State."[241]

To supplement the correct implementation of the measures prescribed in Annex I, the "Commission, acting in cooperation with the appropriate authority of the Member State concerned, shall conduct inspections, including inspections of airports, operators and entities applying aviation security standards, in order to monitor the application by Member States of this Regulation and, as appropriate, to make recommendations to improve aviation security".[242] The detailed rules of the inspection are then contained in

237 Regulation 300/2008, Art. 10.

238 *Ibid.*, Art. 11 and Ann. II. *See*, Commission Regulation (EU) No. 18/2010 of 8 January 2010 Amending Regulation (EC) No. 300/2008 of the European Parliament and of the Council as far as Specifications for National Quality Control Programmes in the Field of Civil Aviation Security are Concerned, OJ L 7, 12 January 2010, pp. 3-14.

239 *Ibid.*, Art. 12.

240 *Ibid.*, Art. 13.

241 *Ibid.*, Art. 14.

242 *Ibid.*, Art. 15(1).

Commission Regulation 72/2010, which lays down the procedures for conducting Commission inspections in the field of aviation security.[243]

As per Article 15(2), the inspections shall be unannounced and only the concerned Member State shall receive prior notification. Pursuant to Article 15(3), a report shall be produced and communicated to the appropriate national authority outlining the measures to be taken to remedy any identified deficiencies. As a result, if and where Regulation 300/2008 is relevant for cybersecurity, there are measures in place to ensure implementation and effective oversight.

243 Commission Regulation (EU) No. 72/2010 of 26 January 2010 Laying Down Procedures for Conducting Commission Inspections in the Field of Aviation Security, OJ L 23, 27 January 2010, pp. 1-5.

III International Regulation of Aviation Cybersecurity

1 Introduction

It is essential that the Chicago Convention and its Annexes are examined as they are explicitly cited in Regulation 216/2008 and Regulation 2018/1139, as well as in Regulation 300/2008, which were all extensively discussed in Chapter II.

Regulation 2018/1139 makes explicit reference to the Chicago Convention. For example, Article 1(2)(g) of the Regulation, which is also expressed in Article 2(2)(d) of Regulation 216/2008, strives to:

> assist Member States, in the fields covered by this Regulation, in exercising their rights and fulfilling their obligations under the Chicago Convention, by ensuring a common interpretation and a uniform and timely implementation of its provisions, as appropriate.

As a result, the Agency has the competencies to interpret and apply Member States' obligations as found in the Chicago Convention, within the scope of the Regulation. Furthermore, Recital 15 of Regulation 2018/1139 supports consistency between the Chicago Convention and EU rules. Therefore, the EU rules can affect how cybersecurity obligations found in the Chicago Convention are applied at a Member State level.

This is also applicable within the EU security regime as Article 1(1) of Regulation 300/2008 "provides the basis for a common interpretation of Annex 17 to the Chicago Convention". As a result, there is a direct relationship between EU law and the Chicago Convention as the former provides a basis for interpretation of the latter.

The Commission and Agency should, in accordance with Article 4(1)(h) of Regulation 2018/1139, "take into account the international rights and obligations in the field of civil aviation of the Union and of the Member States, including those under the Chicago Convention" when taking measures pursuant to the Regulation. This is particularly stressed within the context of implementing measures as, for example, the Commission shall take due account of the Annexes to the Chicago Convention.[1] In contrast to Article

1 *See*, for information on Implementing measures, Regulation 2018/1139, Art. 16a.

1(2)(e), Article 4(1)(h) provides that the Chicago Convention has a direct impact on Agency measures. For example, the product of RMT.0720, namely a Commission Implementing Regulation, will strive to be compatible with the Chicago Convention. Thus, there is an additional relationship between EU law and the Chicago Convention as EU law will drafted in a way to strive for compatibility.

As a result of the relationship between EU law and the Chicago Convention, where each affects the other, Chapter III will explore the relevant areas of the Chicago Convention with the aim of providing a better understanding of the rights and obligations in EU law within the scope of aviation cybersecurity.

2 Setting the Scene: Declaration on Cybersecurity in Civil Aviation

The Declaration on Cybersecurity in Civil Aviation, more commonly referred to as the 'Dubai Declaration', was presented on the 5 April 2017 at ICAO's inaugural Cyber Summit and Exhibition in Dubai, UAE, which was held between 4 and 6 April 2017. While the Dubai Declaration is a non-binding text, it "signifies the commitment and unity of the air transport sector towards achieving effective cyber resilience".[2] Within the text, the Dubai Declaration makes six specific declarations.[3]

Declaration 1 correctly identifies the role of States in mitigating cyber threats, building upon capacities and constructing an appropriate legal framework. It, therefore, places States as the lead responsible stakeholder "to mitigate the risk posed by cyber threats, to build their capability and capacity to address such threats in civil aviation, and to ensure their legislative framework is appropriately established to take action against actors of cyber-attacks".

The role of other stakeholders should not be neglected. Therefore, Declaration 3 stresses the need for collaboration and exchange between States and relevant stakeholders "for the development of an effective and coordinated global framework to address the challenges of cybersecurity in civil aviation". It should be noted that Declaration 3 focuses on vertical collaboration and exchange (State-stakeholder) and does not promote horizontal colla-

2 ICAO, 'ICAO Summit Formalizes New Dubai Declaration to Ensure Global Aviation Remains United, Proactive on Cybersecurity Preparedness', https://www.icao.int/Newsroom/Pages/ICAO-Summit-formalizes-new-Dubai-Declaration-ensures-global-aviation-remains-united-proactive-on-cybersecurity-preparedness.aspx.

3 All quotes found in Chapter III. 5.3 are taken from the Dubai Declaration. ICAO, 'Declaration on Cybersecurity in Civil Aviation', Dubai, UAE, 4-6 April 2017, https://www.icao.int/Meetings/CYBER2017/Documents/Draft%20Dubai%20DECLARATION%20ON%20CYBERSECURITY%20IN%20CIVIL%20AVIATION_10%20March%202017.pdf.

boration and exchange (stakeholder-stakeholder). As a result, the scope of this Declaration 3 is limited.

Increased stakeholder involvement is supported by Declaration 4, which declares that "[c] ybersecurity matters must be fully considered and coordinated across all relevant disciplines within State aviation authorities". Aviation cybersecurity is not just a matter for traditional aviation stakeholders, therefore, Declaration 4 correctly identifies the need to involve stakeholders from other relevant sectors.

Declaration 2 states that "[c]yber capabilities applied to aviation should be used exclusively for peaceful purposes and only for the benefit of improving safety, efficiency and security". As a result, there is an appreciation of the different attackers, motives and results of cyber activities. In this regard, cyber activities that promote safety, efficiency and security, such as 'White Hat' attackers are acceptable. On the other hand, 'Black Hat' attackers are considered unacceptable. This is axiomatic, but the possibility of unacceptable behaviour taking place and the lessons that can be learnt from these activities should not be ignored. However, it is unclear what is meant by the term 'for peaceful purposes', whether this is to be taken from other areas of public international law like the Treaty on Principles Governing the Activities of States in the Exploration and Use of Outer Space,[4] or it is to be understood within the specific context of cybersecurity so passive attack methods are taken into consideration.

Declarations 5 and 6 stick closely to the traditional thinking of aviation security as they link cyberattacks to certain offences. First, Declaration 5 declares that cyberattacks initiated "against civil aviation must be considered an offence against the principles and arrangement for the safe and orderly development of the international civil aviation". In order for this to be realised, there must be law that restricts such acts. This may be found in Article 4 of the Chicago Convention, as discussed in Chapter III.3.3, as cyberattacks may be deemed 'inconsistent with the aims' of the Chicago Convention. However, read in conjunction with Declaration 6 the focus appears to be elsewhere. This is because Declaration 6 promotes the ratification and entry into force of the Beijing Convention and Beijing Protocol as it argues that their application "would ensure that a cyber-attack on international civil aviation is considered an offence" and "would serve as an important deterrent against activities that compromise aviation safety by exploiting cyber vulnerabilities". The effectiveness of the two Beijing instruments in achieving these two goals is debatable as discussed in Chapter III.7.2, but it would allow for harmonised international standards.

4 Treaty on Principles Governing the Activities of States in the Exploration and Use of Outer Space, Including the Moon and Other Celestial Bodies, 610 UNTS 205, 18 UST 2410, TIAS No. 6347, 6 ILM 386, 27 January 1967, Art IV.

3 Convention on International Civil Aviation of 1944

3.1 *Applicability of the Chicago Convention*

The Chicago Convention was signed on 7 December 1944 by 52 States. It is regarded as one of the most successful international treaties as it has 193 Signatory States (as per July 2019[5]) and provides "an overarching and underpinning legal framework to international civil aviation and has been referred to as the Magna Carta of international aviation".[6] While the Chicago Convention is constitutional in nature and is over half a century old, it was drafted in such a way, with clear forward-looking aims and objectives, that it provides the ability for it to be flexible and evolve via its annexes, and to have relevance for cybersecurity. However, before assessing its relevance to cybersecurity, its scope and limitations must first be identified to ascertain its general applicability to aviation generally and aviation cybersecurity more specifically.

3.1.1 International

The Chicago Convention regulates international aviation, which denotes the operation of services that pass through the airspace of more than one contracting State.[7] Article 96 of the Chicago Convention defines 'international air service' as "an air service which passes through the airspace over the territory of more than one State". If this element is not satisfied, then the transit will be considered a domestic flight and regulated outside the scope of the Chicago Convention at a national level by the State Parties or, where such responsibilities are delegated, by regional aviation authorities like the EU.

Article 96, while providing some clarity as to what international means, does not provide a definition of 'State' so it is incomplete. Article 1 on Sovereignty provides some clarity to this issue.

> The contracting States recognize that every State has complete and exclusive sovereignty over the airspace above its territory.

It is, therefore, clear that the Convention is using the term 'State' in an inclusive way to include all States and not just the narrower term 'Contracting State'.

---

5 *See*, for the State Parties of the air law treaties referred to in the work, ICAO, 'Current Lists of Parties to Multilateral Air Law Treaties', https://www.icao.int/secretariat/legal/lists/current%20lists%20of%20parties/allitems.aspx.

6 P. Mendes de Leon and B. Scott, 'An Analysis of Unmanned Aircraft Systems under Air Law', *in* A. Završnik, *Drones & Unmanned Aerial Systems: Legal & Social Implications for Security & Surveillance*, (Springer, 2016), at 187.

7 *See*, M. Milde, *International Air Law and ICAO*, Third Edition, (Eleven, 2016) at 35-49.

Article 1 is a "[r]ecognition by States of their sovereignty over their airspace inevitably presupposes that this rule has already been entrenched in the annals of air law in an earlier instrument".[8] This is a reference to the Convention Relating to the Regulation of Aerial Navigation of 1919 (Paris Convention 1919) which stated "High Contracting Parties recognise that every Power has complete and exclusive sovereignty over the airspace above".[9] This is a recognition of an existing rule where one must look to general public international law. In practice this has little consequence due to the large number of State Parties, with only the United Nation's (UN) States not being party to the Chicago Convention are Dominica and Lichtenstein.

In practice, there needs to be a relationship between two or more States in order for the international element to be satisfied. As Articles 5 and 6 of the Chicago Convention demonstrate, this incorporates flights where an aircraft of one State flies over or lands in another State. Furthermore, cabotage flights are also considered international flights pursuant to Article 7 of the Chicago Convention.

Cabotage: A flight between the same point or two different points within the same State by an operator of a different State. For example, a flight between Cologne and Berlin – both in Germany – by Ryanair, an Irish operator.

The physical scope of the State is limited to its territory which is defined in Article 2 of the Chicago Convention.

> For the purposes of this Convention the territory of a State shall be deemed to be the land areas and territorial waters adjacent thereto under the sovereignty, suzerainty, protection or mandate of such State.

A definition of 'territorial waters' can be found in the UN Convention on the Law of the Sea which declares that States have "the right to establish the breadth of its territorial sea up to a limit not exceeding 12 nautical miles".[10] A flight within a State and outside of its territorial waters that does not interact with another State will not be considered an international flight.[11]

8 R. Abeyratne, *Convention on International Civil Aviation*, First Edition, (Springer, 2014), at 15.

9 Convention Relating to the Regulation of Aerial Navigation of 1919 (Paris Convention), Paris, France, 13 October 1919. *See*, for a detailed discussion on the Paris Convention 1919, John Cobb Cooper, 'The International Air Navigation Conference, Paris 1910', 19 *Journal of Air Law and Commerce* 127-143 (1952).

10 United Nations Convention on the Law of the Sea, 1833 UNTS 3; 21 ILM 1261 (1982), Art 3.

11 Flights over the high seas will be subject to Ann. 2 to the Chicago Convention, Rules of the Air, Tenth Edition, (2005).

The international element required by the Chicago Convention is connected to the aircraft, for example, as observed in Article 15 on airports where the airport's location is static but the aircraft it serves are not. The focus is on the aircraft's connection with foreign States, so this must also be the focus for cybersecurity. While a cyberattack against an aviation stakeholder could come from anywhere in the world, this point is not relevant for the invocation of the Chicago Convention. In other words, a domestic flight which is subject to a cyberattack originating from another State's territory, would not fall under the scope of the Chicago Convention.

Cybersecurity does not produce any new issues for this requirement.

3.1.2 Aviation

The Chicago Convention applies to aviation. However, no definition of the term 'aviation' is provided in the text of the Convention and discussions about the scope of the Convention often focus on the definition of 'aircraft'.

On 6 November 1967, ICAO provided a standard definition of the term 'aircraft' in its Annexes.[12]

> Aircraft: "Any machine that can derive support in the atmosphere from the reactions of the air other than the reactions of the air against the earth's surface."

The term 'aircraft' is divided into lighter than air and heavier than air aircraft as highlighted in Figure 1.

12 Ann. 2 to the Chicago Convention, 2005, at 1-2.

Figure 1 Classifications of Aircraft[13]

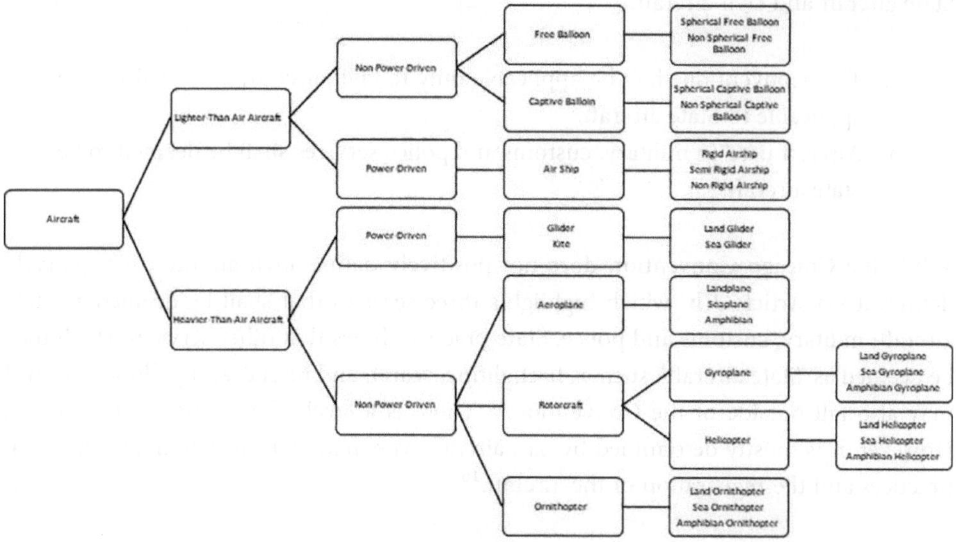

This definition should not be read in a narrow sense so as to mean that the Chicago Convention, and thus its relevance to cybersecurity, is only directly applicable to 'aircraft'. Rather, the Chicago Convention's scope is much wider and should be interpreted to include aircraft and related content. For example, the Convention and its Annexes cover, *inter alia*, aerodromes, radio communications, personnel licencing and air traffic services. Thus, cybersecurity within the context of the Convention is much broader than just the physical aircraft and, therefore, this study will address this wider scope.

The exact scope of the Convention, when placed into cybersecurity, is debatable. For example, it is debatable whether a phishing attack made against an airline's finance department, such as with the Ryanair case,[14] or a DDoS attack against an airline's website that prevents bookings, constitute 'aviation'/'aircraft' within the meaning of the Chicago Convention. While they affect aviation stakeholders, such attacks may stretch the Chicago Convention's scope too far. On the other hand, cyberattacks concerning unauthorised access to an aircraft due to a design vulnerability may be more relevant to the Chicago Convention via the certification of airworthiness requirements under Article 31 and Annex 8. Therefore, a balance needs to be found on the scope of 'aviation'/'aircraft' by States and ICAO.

---

13 Ann. 7 to the Chicago Convention, Aircraft Nationality and Registration Marks, Sixth Edition, July 2012, at 2.

14 *See*, Chapter I.1.

3.1.3 Civil

The Chicago Convention, pursuant to Article 3 (a) and (b), draws a demarcation between State aircraft and civil aircraft.

a. This Convention shall be applicable only to civil aircraft, and shall not be applicable to State aircraft.
b. Aircraft used in military, customs and police services shall be deemed to be State aircraft.

While the Chicago Convention does not positively define 'civil aircraft', it negatively defines it via Article 3(b) which highlights three services that shall be deemed as State aircraft; military, customs and police. State practice shows that other services which may be deemed as 'State aircraft', such as firefighting, search and rescue, and pollution control may also fall outside of the Convention.[15] There is a level of discretion left to States. However, it is mostly determined by use, aircraft type, national legislation, international practices and the registration of the aircraft.[16]

The scope of the Chicago Convention to cybersecurity is limited to cyberattacks on civil aviation only and the cybersecurity measures applied to counter such attacks on civil aviation. However, the scope also includes State cyber activities against civil aviation, in the same way that the shooting down of civilian aircraft by military aircraft, as dealt with under Article 3 *bis*,[17] still falls within the scope of the Chicago Convention.[18]

State aircraft are then dealt with under separate sets of law, notably national legislation and specific international agreements.

3.1.4 War and Emergency Conditions

While the Chicago Convention applies to international civil aviation, Article 89 provides that in cases of war or national emergency, whereby in the latter case notification is made to the Council, the provisions of the Chicago Convention shall not apply.[19] This could include the entirety of the Convention or just specific articles. This happened, for example, on 28 November 1962 when India informed ICAO that "a state of grave emergency

15 *See*, R. de Oliveira, 'The Distinction Between Civil and State Aircraft: Does the Current Legal Framework Provide Sufficient Clarity of Law with Regard to Civil and State Aircraft in Relation to Aviation Practicalities?', 41(4/5) *Air and Space Law* 329-343 (2016).

16 P. Mendes de Leon, 2017, at 14-15.

17 *See*, Chapter III.3.2.

18 *See*, for more information, S. Kaiser, 'Legal Considerations about the Loss of Malaysia Airlines Flight MH17 in Eastern Ukraine', 40(2) *Air and Space Law* 107-121 (2015), at 107.

19 J. Huang, Aviation Safety Through the Rule of Law: ICAO's Mechanisms and Practices, First Edition, (Wolters Kluwer, 2009), at 92.

existed and the Government of India may not find it possible to comply with any or all provisions of the Convention".[20] Article 89 has also been invoked for reasons of 'war' as, for example, Egypt wrote that Israeli aircraft may not fly over Egypt airspace due to a "state of war existing since 14 May 1948, the date of outbreak of hostilities in Palestine".[21] No State has yet invoked this for a cyberattack, however, this may happen in the future when aviation stakeholders are seriously targeted.

Article 35 of the Chicago Convention is also relevant for 'war' as it prohibits the carriage of cargo.

> a. No munitions of war or implements of war may be carried in or above the territory of a State in aircraft engaged in international navigation, except by permission of such State. Each State shall determine by regulations what constitutes munitions of war or implements of war for the purposes of this Article, giving due consideration, for the purposes of uniformity, to such recommendations as the International Civil Aviation Organization may from time to time make.

A State may wish to team certain technologies that have cyberattack capabilities within the wording of this article so as to prohibit its distribution and reduce the risk to aviation stakeholders.

3.1.5 Cybersecurity

If the above criteria are satisfied, then the Chicago Convention is applicable to that aviation activity. The term 'cyber' does not feature in the text of the Chicago Convention, so it may seem quiet on the topic. However, the next step is to look at the specific Articles and Annexes of the Chicago Convention to assess their relevance to cybersecurity. As with EU law, the rules may either directly or indirectly regulate cybersecurity, or become relevant in the future with technological advancement, such as new attack methods or new cybersecurity features on aviation technology, and regulatory developments, such as amendments to Annexes. This subchapter, having already dealt with Articles 1, 2 and 3 of the Chicago Convention, and will begin the analysis with Article 3 *bis* and then advance to other articles to show direct and indirect relevance.

---

20 ICAO Doc. 8900/2. Repertory-Guide to the Convention on International Civil Aviation, Second Edition (1977), at 1-2.

21 *Ibid.*

3.2 *Shooting Down of Civilian Aircraft*

3.2.1 Setting the Scene

Since the adoption of the Chicago Convention there have been three amendments that have introduced new Articles to the text.

1. Article 3 *bis*: Non-use of weapons against civil aircraft in flight.[22]
2. Article 83 *bis*: Transfer of certain functions and duties.[23]
3. Article 93 *bis*: Expulsion from ICAO or suspension of membership.[24]

This Section will focus on Article 3 *bis* and whether its scope, through the use of the term 'weapons', applies to different attack methods.

There are many examples of 'weapons' being used against civilian aircraft.[25] An early example is the case of the Libyan aircraft that was shot down with a missile by Israel in 1973 resulting in the tragic deaths of 106 people on board.[26] However, it was the shooting down of Korean Air Lines Flight 007 in 1983 that provoked the international community to amend the Chicago Convention with a specific Article. This flight was a scheduled Korean Air Lines flight from New York City, US to Seoul, South Korea via Anchorage, US. On 1 September 1983, the aircraft was shot down by a Soviet Su-15 interceptor aircraft, when, en-route from Anchorage to Seoul, it deviated from its flight path and flew through the prohibited Soviet airspace. A more recent case of a weapon being used against civilian aircraft happened on 7 July 2014, when Malaysia Airlines flight MH17 from Amsterdam, The Netherlands to Kuala Lumpur, Malaysia was shot down by a Buk surface-to-air missile over Ukraine at an altitude of 33,000 ft, tragically killing all 298 persons on board.[27]

The reaction of the international community to Flight 007 was to amend the Chicago Convention with the inclusion of Article 3 *bis*, which was adopted by consensus on

22 Protocol Relating to an Amendment to the Convention on International Civil Aviation [Art. 3 *bis*], 1 October 1998.

23 Protocol Relating to an Amendment to the Convention on International Civil Aviation [Art. 83 *bis*], 20 June 1997.

24 Protocol Relating to an Amendment to the Convention on International Civil Aviation [Art. 93 *bis*], 20 March 1961.

25 *See*, for historical overview, B. Foont, 'Shooting Down Civilian Aircraft: Is There an International Law', 72 *Journal of Air Law and Commerce* 695-725 (2007).

26 *See*, for more information, Res. A 19-1: Shooting Down of a Libyan Civil Aircraft by Israeli Fighters on 21 February 1973, *in* ICAO, Assembly Resolutions in Force (as of 8 October 2004), Doc. 9848, at I-27.

27 *See*, S. Kaiser, 2015.

10 May 1984, less than one year on from Flight 007, by the 25th Session (Extraordinary) of the Assembly.[28] Article 3 *bis* (a) provides:

> The contracting States recognize that every State must refrain from resorting to the use of weapons against civil aircraft in flight and that, in case of interception, the lives of persons on board and the safety of aircraft must not be endangered. This provision shall not be interpreted as modifying in any way the rights and obligations of States set forth in the Charter of the United Nations.[29]

The Article, as an amendment, requires adoption by States. It entered into force on 1 October 1998 and has 156 State Parties (as of July 2019). As a result, the amendment is not binding on those Chicago Convention contracting States that have yet to adopt Article 3 *bis*.

Article 3 *bis* uses the term 'recognise', so it has been argued by some commentators that this is "not a new rule of law", but the recognition of "the existence of a prior rule binding on all parties and prohibiting the use of weapons against civil aircraft in flight".[30] It is simply a codification of other sources of law;[31] the obligation not to shoot down civilian aircraft is found in other sources of international law.[32]

In order for Article 3 *bis* to apply, three criteria must be satisfied whereby:
1. there must be a use of a 'weapon',
2. the use of a weapon must be 'against a civil aircraft', and
3. the aircraft must be 'in flight'.

First, the term 'aircraft' has been sufficiently discussed above in Chapter III.3.1 so no further reflection is required.

28 Protocol Relating to an Amendment to the Convention on International Civil Aviation [Art. 3 *bis*], 10 May 1984.

29 *See*, for a general overview of Art. 3 *bis*, R. Abeyratne, 2014, at 67-73.

30 G. Guillaume, 'The Destruction on 1 September 1983 of the Korean Airlines Boeing (Flight KE 007)', ITA Magazine No. 0-18 (1984), at 34, *in* J. Huang, 2009, at 84-88.

31 *See*, M. Milde, 'Interception of Civil Aircraft vs. Misuse of Civil Aviation (Background of Amendment 27 to Annex 2)', XI *Annals of Air and Space Law* 105 (1986), at 125.

32 Article 3 *bis* is declaratory of the existing general international law with respect to the following elements:
a. the obligation of States to refrain from resorting to the use of weapons against civil aircraft in flight;
b. the obligation, in case of interception, not to endanger the lives of persons on board and the safety of aircraft;
c. the right of States to require landing at a designated airport of a civil aircraft flying above its territory without authority or if there are reasonable grounds to conclude that it is being used for any purpose inconsistent with the aims of the Convention.

ICAO, 'Administrative Package for Ratification of the Protocol on Article 3 *Bis*', https://www.icao.int/secretariat/legal/Administrative%20Packages/3bis_en.pdf.

Second, the definition of 'in flight' is open for debate and has been extensively discussed in international criminal air law.[33] This term could look at the facts; the aircraft is in flight while it is off the ground relying on aerodynamic lift. Alternatively, it could be read in line with other aviation treaties, for example, the Tokyo Convention in Article 1(3) provides a different definition for the term 'in flight'.

> For the purposes of this Convention, an aircraft is considered to be in flight from the moment when power is applied for the purpose of takeoff until the moment when the landing run ends.

Article 5(2) of the Tokyo Convention also provides a definition of 'in flight' within the scope of the aircraft commander's powers.

> [A]n aircraft shall for the purposes of this Chapter, be considered to be in flight at any time from the moment when all its external doors are closed following embarkation until the moment when any such door is opened for disembarkation. In the case of a forced landing, the provisions of this Chapter shall continue to apply with respect to offences and acts committed on board until competent authorities of a State take over the responsibility for the aircraft and for the persons and property on board.

Thus, this has a wider scope than the other definition found in the same Convention, allowing it to apply even during acts of sabotage on the ground. Article 2(b) of the Convention for the Suppression of Unlawful Acts against the Safety of Civil Aviation (Montreal Convention 1971) uses the alternative term 'in service'.[34]

> [A]n aircraft is considered to be in service from the beginning of the preflight preparation of the aircraft by ground personnel or by the crew for a specific flight until twenty-four hours after any landing; the period of service shall, in any event, extend for the entire period during which the aircraft is in flight as defined in paragraph (a) of this Article.

The duration of in flight is then open for debate with various Treaties providing interpretive value. However, as this is a general issue of interpretation, providing nothing specific to cybersecurity, it will not be dealt with any further.

---

33 *See*, Chapter III.7.

34 Convention for the Suppression of Unlawful Acts against the Safety of Civil Aviation (Montreal Convention 1971), 974 UNTS 177, 24 UST 564, 10 ILM 1151, 23 September 1971.

Third, the event must involve a 'weapon'. This has direct relevance for cybersecurity, as a certain attack method or all attack methods may constitute a weapon pursuant to Article 3 *bis*. Therefore, this will be discussed in more detail as it may be interpreted to include some types of cyberattacks.[35]

3.2.2 Background to Article 3 bis

The Chicago Convention does not make any attempt to define the term 'weapon' and neither do any of the Annexes or Treaties concluded under the auspices of ICAO, like the criminal Treaties discussed in Chapter III.7.

Each of the incidents concerning the shooting down of civil aircraft by missiles mentioned above has triggered heated debates. In the case of the Korean Air Lines Flight 007, when the Security Council of the UN was not able to take action due to the Soviet Union's veto, the ICAO Council was able to adopt a resolution on 6 March 1984 which "[c]ondemns the use of armed force which resulted in the destruction of the Korean airliner and the tragic loss of 269 lives".[36] Here, the term 'use of armed force' was chosen, which has its roots in public international law, such as Article 2(4) of the UN Charter.[37]

The Chicago Convention, as affirmed by Article 3 *bis*, is a part of the wider international law context. Notably, this is Article 2(4) of the UN Charter, which generally prohibits the use of force except in the case of self-defence as set out in Article 51 or with Security Council authorisation via Article 39 of the UN Charter. The general discussions on the definition of 'use of force' as found in Article 2(4) of the UN Charter would be appropriate to assess the scope of Article 3 *bis*, however, this term was not utilised.

With respect to the Iran Air incident on 3 July 1988, the Iranian observer stated that an "act of violence, irrespective of who or which country the perpetrator might be, will not be left unaccounted".[38] Subsequently, the UK presented a draft resolution on behalf of nine States, however, this did not satisfy Iran.[39] The Soviet Union and Czechoslovakia then proposed to amend the sixth clause of the draft.

> CONDEMNS the use of armed force against civil aviation, including the act which resulted in the tragic destruction of an Iran Air airliner and the loss of

35 *See*, for more information on treaty interpretation, Vienna Convention on the Law of Treaties, United Nations, Treaty Series, Vol. 1155, p. 331, 23 May 1969, Art. 31.

36 *See*, J. Huang, 2009, at 89.

37 Charter of the United Nations, 1 UNTS XVI, 24 October 1945.

38 J. Huang, 2009, at 89.

39 ICAO Council Working Paper, 'Draft Resolution', C-WP/8821, 16 March 1989. This document was presented by Australia, Canada, France, the Federal Republic of Germany, Italy, Japan, Spain, Switzerland and the United Kingdom.

> 290 lives, while noting the accidental sequence of events and errors in the identification of the aircraft.[40]

The phrase 'acts of violence' is not unknown to international civil air law, as it can be found in the Tokyo Convention of 1968 Article 1(1) and the Montreal Protocol 1988 Article 1 *bis*(a). While the context is different, its interpretation could serve as an interpretative tool. However, the US was opposed to the amendment proposed by the Soviet Union and Czechoslovakia. This was because it concluded that the specifics of each case are different, thus it cannot be the case that "all uses of force against civil aircraft in flight deserved equal condemnation, without due regard to the particular facts and circumstances surrounding individual cases".[41]

In the final adopted text, the focus is on the condemnation of 'the use of weapons'. This term is not used in any source of air law, therefore, interpretative analysis must look beyond air law.

3.2.3 Definition in International Law

There is currently no definition of 'weapon' in international law, rather the extensive body of international law related to weapons opts to create different categories of weapons. There is currently no categorisation of 'cyberweapon', but Protocol IV on Blinding Laser Weapons is of interpretative use.

> It is prohibited to employ laser weapons specifically designed, as their sole combat function or as one of their combat functions, to cause permanent blindness to unenhanced vision, that is to the naked eye or to the eye with corrective eyesight devices. The High Contracting Parties shall not transfer such weapons to any State or non-State entity.[42]

Protocol IV provides two interesting points for the discussion of a cyberattack as a weapon.

1. A specific attack method is categorised as a weapon, rather than defining the overarching term 'weapon'.
2. Weapons do not have to be tangible, rather they can also be electromagnetic radiation, such as lasers.

40 J. Huang, 2009, at 90.
41 *Ibid.*
42 Protocol on Blinding Laser Weapons (Protocol IV), Vienna, 13 October 1995, Art. 1.

As a result, it should not be precluded that a cyberattack could be classified in the future as a type of weapon under this international law regime. Electromagnetic radiation could include electronic-based cyberattack methods, such as through communication and navigation links.

Protocol I allows for new weapons to be considered in the international legal regime under Article 36.[43]

> In the study, development, acquisition or adoption of a new weapon, means or method of warfare, a High Contracting Party is under an obligation to determine whether its employment would, in some or all circumstances, be prohibited by this Protocol or by any other rule of international law applicable to the High Contracting Party.[44]

This has yet to be done for cyberattacks, but this has not stopped calls for such an approach. For example, Google has proposed a Digital Geneva Convention "that will call on the world's governments to pledge that they will not engage in cyberattacks on the private sector, that they will not target civilian infrastructure, whether it's of the electrical or the economic or the political variety".[45]

Additionally, China and Russia both submitted a draft treaty in February 2008 to the Conference on Disarmament on the prohibition of the placement of any weapon in outer space. The proposal has a relevant definition.

> c) [T]he term 'weapons in outer space' means any device placed in outer space, based on any physical principle, specially produced or converted to eliminate, damage or disrupt normal function of objects in outer space, on the Earth or in its air, as well as to eliminate population, components of biosphere critical to human existence or inflict damage to them.[46]

---

43 *See*, International Committee of the Red Cross, 'A Guide to the Legal Review of New Weapons, Means and Methods of Warfare: Measures to Implement Article 36 of Additional Protocol I of 1977 International Committee of the Red Cross Geneva', January 2006, Reports and Documents, Vol. 88, Number 864, December 2006.

44 Protocol Additional to the Geneva Conventions of 12 August 1949, and relating to the Protection of Victims of International Armed Conflicts (Protocol I), 8 June 1977, Art. 36.

45 *See*, Microsoft, The Need for a Digital Geneva Convention, https://www.blogs.microsoft.com/on-the-issues/2017/02/14/need-digital-geneva-convention/.

46 Draft Treaty on Prevention of the Placement of Weapons in Outer Space and of the Threat or Use of Force Against Outer Space Objects, Russia-China, 29 February 2008, Art. I(c).

While this is only a draft treaty, thus lacking legal force, it shows an international attempt to define 'weapon'. Here it must be: (1) a device that (2) has some sort of negative outcome.

Despite the limited use that these definitions found in public international law have for interpreting Article 3 *bis*, they do present a theme whereby the weapon must have some kind of an effect on a third party. As such, if a cyberattack causes or could cause injury, death, damage or destruction, then this would correlate to the above definitions. Without such, it is unlikely that the cyberattack will be a weapon for the purpose of Article 3 *bis*.

3.2.4 Definition in 'Cyber Law'

The European Convention on Cybercrime is a specific treaty regulating States' behaviour in the context of cybersecurity (*lex specialis*). It obliges State Parties to criminalise certain cyber activities by individuals, does not define the term 'weapon'.[47]

The Tallinn Manual addresses the question as to what constitutes a cyber weapon in Rule 41(2), where it indicates that:

> [...] cyber weapons are 'cyber means of warfare' that are by design, use, or intended use capable of causing either (i) injury to, or death of, persons; or (ii) damage to, or destruction of, objects, that is, causing the consequences required for qualification of a cyber operation as an attack.

This definition includes all means that could be used for damage, regardless of intent to cause damage. As a result, this would include all cyberattacks that cause injury, death, damage or destruction. This could include any cyberattack beyond espionage or manipulation of data in terms of interference.[48] This definition is limited as the Tallinn Manual is a non-binding academic study on how international law applies to cyber conflicts and cyber warfare. While useful, it must be stressed that this is not a source of law and that Article 3 *bis* does not use the term 'cyber weapon'.[49]

3.2.5 Closer to a Definition?

A definition of the term 'weapon' is missing from international law.

---

47 *See*, D. Blake and J. Imburgia, "Bloodless Weapons'? The Need to Conduct Legal Reviews of Certain Capabilities and the Implications of Defining them as 'Weapons", 66 *Air Force Law Review* 157-203 (2010). *See*, Chapter 7.4.

48 G. Brown and A. Metcalf, 'Easier Said Than Done: Legal Reviews of Cyber Weapons', 7(1) *Journal of National Security Law and Policy* 115-138 (2014), at 135.

49 Statute of the International Court of Justice, 3 Bevans 1179; 59 Stat. 1031; T.S. 993; 39 AJIL Supp. 215, 24 October 1945, Art. 38(1)(d).

> The term 'weapons' has not been defined but its interpretation in practice should not cause any difficulties when the ordinary meaning of the word is considered; it would include guns, machine guns, cannons, missiles or targeted drones or any other object or substance that can damage or destroy an aircraft in flight.[50]

While this ordinary meaning may be clear for conventional weapons, such as missiles, this is not the case for cyberattacks.

Looking at similar terms is only of limited value. The approach taken from discussions on the 'use of force' and 'cyber weapons' attempt to place an object inside or outside the scope of 'weapon'. However, this is not what Article 3 *bis* asks. This is not the purpose of Article 3 *bis* as it is neither concerned with the specifics of the attack method nor with how the attack method was implemented against a civil aircraft in flight. Rather, what is relevant is the expected or actual outcome. Therefore, the classification of weapon should be *de facto*, rather than *de jure*. In other words, if a cyberattack causes injury, death, damage or destruction, then this would constitute a weapon for the purpose of Article 3 *bis*. However, where this is not the case, such as with passive attacks, such cyberattacks would not amount to a use of a weapon.

It is also useful not to pursue a definition of 'weapon', especially within the context of cyberattacks because "[n]ot only is software intangible, it is subject to frequent changes; different versions of software can bear little resemblance to each other".[51] It could, thus, result in a situation where the development of technology results in the definition becoming obsolete and not covering all situations.

3.3 *Misuse of Civil Aviation*

Article 4 of the Chicago Convention declares that "[e]ach contracting State agrees not to use civil aviation for any purpose inconsistent with the aims of this Convention". The question, therefore, arises as to the extent to which this covers a cyberattack.

The Preamble provides guidance on the aims of the Chicago Convention.

> Whereas the future development of international civil aviation can greatly help to create and preserve friendship and understanding among the nations and

50 M. Milde, 2016, at 61.
51 G. Brown and A. Metcalf, 2014, at 133.

> peoples of the world, yet its abuse can become a threat to the general security and;
>
> Whereas it is desirable to avoid friction and to that cooperation between nations and peoples upon which the peace of the world depends;
>
> Therefore, the undersigned Governments having agreed on certain principles and arrangements in order that international civil aviation may be developed in a safe and orderly manner and that international air transport services may be established on the basis of equality of opportunity and operated soundly and economically.[52]

The Preamble, in providing interpretative guidance, shows that the aim is to facilitate international cooperation for the pursuance of global peace and security, and furtherance of aviation safety.

Cuba provided a practical example of such acts deemed inconsistent with the aim of the Chicago Convention during the 25th (Extraordinary) Session of the Assembly in Montreal between 24 April and 10 May 1984.

> Acts of aggression, infiltration or espionage involving discharge of harmful substances or pathogenic agents; transport of contraband or prohibited traffic using the airspace of another State, even with destination to a third State or with any other purpose inconsistent with the aims of the Convention.[53]

Some examples, such as the shooting down of civilian aircraft and the destruction of Gaza International Airport, have been interpreted as acts that are inconsistent with the aims of the Chicago Convention.[54]

According to the ICAO Study on Article 4 conducted by the Secretary General, "an inconsistent purpose with the aims of this Convention" has to be interpreted to mean "threats to the general security".[55] Therefore, such acts by States that threaten general security, which could include cybersecurity, through the use of aviation would come under the scrutiny of Article 4. This would seem to be the case with the use of a cyber-

52 Chicago Convention, Preamble.
53 A25-Min. Ex/6, *in* R. Abeyratne, 2014, at 75.
54 *Ibid.*, at 75-94.
55 ICAO Council, "Misuse' of Civil Aviation', 117th Session, C-WP/8217, 18 March 1986.

attack within the terms of Article 3 *bis* and, following Cuba's definition, passive attacks, such as cyber espionage like eavesdropping and tapping.[56]

Article 4 has limited application for cyberattacks as it is focused on State to State relations and would be dealt with under the dispute settlement measurements found in Chapter XVIII of the Chicago Convention. As a result, cyberattacks between non-State actors will not fall under the scope of this Article.

3.4 *Scheduled and Non-scheduled Air Services*

3.4.1 Introductory Remarks

Article 1 of the Chicago Convention confirmed the customary international law principle that a State has sovereignty over its air space.[57] Thus, it is clear that civil aircraft have no automatic right to fly over or land in the territory of another State, which is also applicable, as affirmed in Article 3(c), to State aircraft. Consequently, the Chicago Convention had to provide a legal mechanism to allow aircraft from one State to enter the airspace of another State, otherwise international civil aviation would not be possible. This was provided for under Articles 5, 6, 7 and 8 of the Chicago Convention. This subchapter will be used to demonstrate how secondary law, which stems from obligations found in the Chicago Convention, may be relevant for cybersecurity.

3.4.2 Non-Scheduled Air Services

The first mechanism is Article 5, which permits non-scheduled air services to exercise transit and traffic rights in the territory of another State without the need to obtain prior permission.[58] ICAO defined a non-scheduled air service as "a commercial air transport service performed as other than a scheduled air service"[59], which can include private business jets and charter operations. While this mechanism appears to be liberal, requiring open access, in practice it is more limited.

---

56 *See*, Chapter I.2.2.

57 P. Mendes de Leon, 2017, at 9-10.

58 Chicago Convention, Art. 5:

> Each contracting State agrees that all aircraft of the other contracting States, being aircraft not engaged in scheduled international air services shall have the right, subject to the observance of the terms of this Convention, to make flights into or in transit non-stop across its territory and to make stops for non-traffic purposes without the necessity of obtaining prior permission, and subject to the right of the State flown over to require landing. Each contracting State nevertheless reserves the right, for reasons of safety of flight, to require aircraft desiring to proceed over regions which are inaccessible or without adequate air navigation facilities to follow prescribed routes, or to obtain special permission for such flights.

59 ICAO, Tenth Session of the Statistics Division, 'Review of the Classification and Definitions used for Civil Aviation Activities', STA/10-WP/7 (2009), App. B.

> The over-flown State can, in the interest of safety, require that non-scheduled aircraft obtain special (prior) permission for such flights or follow prescribed routes over areas deemed inaccessible or without adequate air navigation facilities. This effectively means the aircraft operator can also be required to inform the over-flown State of their flight plan, intended flight path and technical stop-over sites, and furnish other information before entering foreign airspace as well.[60]

There is likely to be some kind of communication between the overflown State and/or State disembarkation and the air operator, State of registry and/or State of embarkation. It could, therefore, be possible that such conversations, in the interest of safety and security, may include cybersecurity. However, no evidence to support this assertion could be found.

3.4.3 Scheduled Air Services

In addition to non-scheduled air services, scheduled air services are covered under Article 6.[61] The ICAO Council provides a definition for 'scheduled service':

> A scheduled international air service is a series of flights that possesses all of the following characteristics:
> a. it passes through the airspace over the territory of more than one state;
> b. it is performed by aircraft for the transport of passengers, mail or cargo for remuneration, in such a manner that each flight is open to use by members of the public;
> c. it is operated, so as to serve traffic between the same two or more points; either,
>   i. according to a published time-table, or
>   ii. with flights so regular or frequent that they constitute a recognizable systemic series.[62]

Consequently, prior permission will be required, usually by way of a bilateral or multilaterally air service agreement (ASA), sometimes referred to as 'air transport agreement', in order for a State to exercise transit and traffic rights.

---

60 R. Harding, 'The Law of Air Mobility – The International Legal Principles Behind the US Mobility Air Forces' Mission', 66 *The Air Force Law Review* 1-330 (2010), at 11-12.

61 Chicago Convention, Art. 6:
No scheduled international air service may be operated over or into the territory of a contracting State, except with the special permission or other authorization of that State, and in accordance with the terms of such permission or authorization.

62 ICAO, Definition of Scheduled International Air Service, ICAO Doc. 7278-C/841 (1952).

Bilateral: An ASA between only two States. Multilateral: An ASA between more than two States.

It is possible to generalise air service agreements and group them into different models depending on their restrictions,[63] however, despite this, no two ASAs are the same and because there are currently over 5,000 bilateral ASAs regulating air services worldwide and each must be dealt with individually. Consequently, a case study must be evaluated in order to determine how cybersecurity is treated stemming from Articles 1, 5 and 6 of the Chicago Convention.[64]

3.4.4 Case Study: US-EU Air Transport Agreement

The agreement between the US and EU has been selected as a case study for five reasons:

1. The focus of this study is on the regulation of cybersecurity by the EU, so an EU ASA ought to be assessed.
2. The US-EU ASA is a hybrid of both the US and EU template ASAs, so its content is much wider than just the relationship between these two jurisdictions.[65]
3. It is an Open Skies agreement, so it is an example of the most up-to-date model of ASA.
4. The US and the EU are leaders in aviation safety and security.
5. It is subject to review – having been amended – with the Joint Committee meeting on an annual basis to discuss the ASA. Therefore, it is evolving to combat new issues.

Prior to the current ASA, air transport between the US and EU was governed on a bilateral basis between the US and the individual EU Member States. However, since 2002 following the Open Skies Judgement,[66] EU Member States could "no longer act in isolation when negotiating international air services agreements".[67] The US-EU Agreement

---

63 *See*, as way of comparison, the Bermuda I Agreement which was a bilateral agreement between the US and the UK. It became a template across the globe for bilateral agreements. *See*, Agreement between the government of the UK and the government of the US relating to Air Services between their respective Territories, Treaty Series No. 3, 11 February 1946; and the Open Sky model which is currently the most liberal model. This set-up allows the airlines to organise themselves with very little regulation by the States. An example of such an Agreement is that between the US and Europe. *See*, US-EU Air Transport Agreement, 24 June 2010; and US-EU Air Transport Agreement, 30 April 2007.

64 *See*, Articles 7 and 8 of the Chicago Convention as they are also mechanisms for allowing access to a foreign airspace within specific circumstances (*i.e.* cabotage and unmanned aircraft).

65 The term 'two jurisdictions' refers to the US and the EU, as opposed to the US and the individual EU Member States.

66 Cases C-466/98, C-467/98, C-468/98, C-469/98, C-471/98, C-472/98, C-475/98 and C-476/98 *Commission v. UK, Denmark, Sweden, Finland, Belgium, Luxembourg, Austria, Germany*.

67 Commission, 'External Aviation Policy', https://wwwec.europa.eu/transport/modes/air/international_aviation/external_aviation_policy_en.

came into force on 30 March 2008 marking the end of bilateralism between the US and the EU Member States.[68] This Agreement was the first stage in cooperation as it included a clear roadmap in Article 21 for the negotiations of a Second Stage Agreement. The Protocol to amend the Air Transport Agreement between the US and EU was signed and provisionally applied as from 24 June 2010. Iceland and Norway were also added to the agreements.[69]

Pursuant to Article 3(1) of the First Agreement, the provision grants to the other Party the following rights:

a. the right to fly across its territory without landing;
b. the right to make stops in its territory for non-traffic purposes;
c. the right to perform international air transportation between points on the following routes:
    i. for airlines of the United States (hereinafter 'US airlines'), from points behind the United States via the United States and intermediate points to any point or points in any Member State or States and beyond; and for all cargo service, between any Member State and any point or points (including in any other Member States);
    ii. for airlines of the European Community and its Member States (hereinafter 'Community airlines'), from points behind the Member States via the Member States and intermediate points to any point or points in the United States and beyond; for all cargo service, between the United States and any point or points; and, for combination services, between any point or points in the United States and any point or points in any member of the European Common Aviation Area [...] as of the date of signature of this Agreement; and
d. the rights otherwise specified in this Agreement.

Access to such rights centres on authorisation based on Article 4. If the following are met, then authorisation shall be granted per the Agreement.

68 US-EU Air Transport Agreement, 30 April 2007.
69 US-EU (Iceland, Norway) Air Transport Agreement of 21 June 2011.

a. for a US airline, substantial ownership and effective control of that airline are vested in the United States, US nationals, or both, and the airline is licensed as a US airline and has its principal place of business in US territory;
b. for a Community airline, substantial ownership and effective control of that airline are vested in a Member State or States, nationals of such a state or states, or both, and the airline is licensed as a Community airline and has its principal place of business in the territory of the European Community;
c. the airline is qualified to meet the conditions prescribed under the laws and regulations normally applied to the operation of international air transportation by the Party considering the application or applications; and
d. the provisions set forth in Article 8 (Safety) and Article 9 (Security) are being maintained and administered.

Discussions on authorisation of air service agreements often focus on the ownership and control clause as found in points (a) and (b) of Article 4. However, for cybersecurity points (c) and (d) are of relevance and will instead be analysed.

First, aviation law could provide cybersecurity provisions relevant for the airline, such as those found in the EU safety and security Regulations, as well as in the Annexes to the Chicago Convention. As a result, these cyber-based obligations may be relevant for Point (c). The content of each text must be examined individually to assess the exact relevance of Point (c) to cybersecurity.

Second, Articles 8 on safety and 9 on security may contain cybersecurity requirements, and if they are not met, then pursuant to Article 5(3), the state[70] can "withhold, revoke, limit or impose conditions on the operating authorization or technical permission of an airline or airlines of the other Party in accordance with the provisions of Article 8 (Safety) or Article 9 (Security)".

Article 8 on safety provides very little support with regard to cybersecurity. First, Article 8 (1) provides a mechanism for the mutual recognition of:

> certificates of airworthiness; certificates of competency, and licenses issued or validated by each other and still in force, provided that the requirements for

70 The US-EU ASA uses the term 'Party', however, in the interests of consistency, the term 'state' has been used in this study.

> such certificates or licenses at least equal the minimum standards that may be established pursuant to the Convention.[71]

This does not directly involve cybersecurity as it is concerned with the mutual recognition of certificates and licences, as opposed to the actual issuance of these certificates. Thus, there is only a secondary link to cybersecurity if the national or regional requirements, such as via RMT.0648,[72] contain such cybersecurity requirements for the issuance of certificates.

Article 8(2) allows the responsible authorities of the States to:

> request consultations with other responsible authorities concerning the safety standards maintained by those authorities relating to aeronautical facilities, aircrews, aircraft, and operation of the airlines overseen by those authorities.

Pursuant to Article 8(2), if the requesting responsible authority concludes that the responsible authority of the other State does "not effectively maintain and administer safety standards and requirements in these areas that at least equal the minimum standards that may be established pursuant to the Convention" then, where safety standards are in place based on cybersecurity or where a cyberattack could affect safety, this could form a basis for limiting authorisation. Here, the requesting responsible authorities can withhold, revoke or limit authorisation or technical permission. Again, as with Article 8(1), this is a secondary link to cybersecurity, as opposed to a direct one.

Finally, Article 8(4) confirms that the responsible authorities of the contracting States have the right to conduct safety discussions. This could include cybersecurity issues, though not necessarily. No examples of cybersecurity could be found within the context of Article 8, however, this could change with the persistence of cyberattacks and the growth in regulations.

Article 9, on the other hand, can have direct relevance to cybersecurity. First, Article 9(1) reaffirms that States have an "obligation to each other to protect the security of civil aviation against acts of unlawful interference" and that the States shall act in conformity with the Tokyo Convention 1963, The Hague Convention 1970, the Montreal Convention 1971, and the Montreal Protocol 1988. The importance and relevance of Article 9(1) will be shown in Chapter III.4.3 in its discussions on 'unlawful interference' and Chapter III.7 involving the international criminal conventions.

---

71 The term 'Convention' refers to the Chicago Convention.

72 *See*, Chapter II.7.2.1.

Article 9(2) requires that the States:

> shall provide upon request all necessary assistance to each other to address any threat to the security of civil aviation, including the prevention of acts of unlawful seizure of civil aircraft and other unlawful acts against the safety of such aircraft, of their passengers and crew, and of airports and air navigation facilities.

This assistance provision is very broad as it refers to any threat-based on security. However, the assistance shall only come upon request and to the level that the requesting State deems necessary. As the EU's cyber sharing initiatives do not involve the US government, having a focus on EU stakeholders, it remains to be seen how forthcoming the States will be with regard to such requests based on cybersecurity.

Pursuant to Article 9(3), the Parties shall conform to the aviation security SARPs established by ICAO. As, for example, Annex 17 makes specific reference to 'cyber threats',[73] there is a direct link between the US-EU ASA and cybersecurity in the ICAO Annexes.

A cyberattack can have severe consequences and be implemented in an unforeseen way. Therefore, States have to act pragmatically to the threat. Article 9(6) provides for this as a State "may take emergency measures including amendments to meet a specific security threat". These measures have to be immediately reported to the other Party.

An important step in mitigating a cyberattack is cooperation between potential victims. Article 9(7) underlines the importance of combined efforts to secure harmonisation. Further, Article 9(9) can be used as a tool to facilitate this as, in the event of:

> incident or threat of an incident of unlawful seizure of aircraft or other unlawful acts against the safety of passengers, crew, aircraft, airports or air navigation facilities occurs, the Parties shall assist each other by facilitating communications and other appropriate measures intended to terminate rapidly and safely such incident or threat.

Despite this, Article 9(4) acknowledges the important individual role that each State has with regard to security measures.

If a State has 'reasonable grounds' to believe departure from Article 9 by the other State has occurred, then immediate consultation may be requested per Article 9(10). Where

73 Ann. 17 to the Chicago Convention, 2017, at 4.9.

agreement cannot be reached on this matter, operating authorisation and technical permissions can be withheld, revoked, limited or conditions can be imposed.

The US-EU Agreement does not use the term cybersecurity, so it may *prima facia* appear to be silent on the issue. However, this is not the case as it, for example, makes links to unlawful interference which, as discussed below in Chapter III.4.3, may include cyberattacks. Cybersecurity may be relevant to both Article 8 and 9, however, the link is much stronger with regard to security due to ICAO's Annex 17. Thus, this often-overlooked area of air law is relevant in cybersecurity discussions and may provide a legal mechanism for the enhancement of cybersecurity and the mitigation of cyberattacks. However, this has yet to be tested. Further, the US-EU Agreement is just one in over 5,000 ASA, so its influence is limited. The specifics of each ASA must be assessed to determine the exact impact on cybersecurity. No example of an ASA that directly references cybersecurity could be found.

3.5 *Other Articles of the Chicago Convention*

Chapter III.3 dealt with the applicability of specific Articles of the Chicago Convention – notably Articles 1, 2, 3, 3 *bis*, 4, 5 and 6 – and their direct and indirect relevance for cybersecurity. These have been identified as the Articles that have a link to cybersecurity without reference to Chicago Convention's detailed 19 Annexes. However, other Articles may be relevant indirectly, through the development of related Annexes, or with the development of national or regional law as seen in Chapter II with regard to the EU safety regulations. As a result, these should not be overlooked and their relevance for this topic will be discussed in Chapter III.4 and its examination of the Annexes. However, as no direct relevance could be found, no further Articles will be analysed.

4 Annexes to the Chicago Convention

4.1 Introductory Remarks

There are 19 detailed technical Annexes that were developed and are regularly amended by ICAO pursuant to Article 37 of the Chicago Convention.[74]

> Each contracting State undertakes to collaborate in securing the highest practicable degree of uniformity in regulations, standards, procedures, and organization in relation to aircraft, personnel, airways and auxiliary services in all matters in which such uniformity will facilitate and improve air navigation.

In order to achieve this, ICAO shall, as is deemed necessary, adopt and amend the Annexes within the following areas:

> a. Communications systems and air navigation aids, including ground marking;
> b. Characteristics of airports and landing areas;
> c. Rules of the air and air traffic control practices;
> d. Licensing of operating and mechanical personnel;
> e. Airworthiness of aircraft;
> f. Registration and identification of aircraft;
> g. Collection and exchange of meteorological information;
> h. Log books;
> i. Aeronautical maps and charts;
> j. Customs and immigration procedures;
> k. Aircraft in distress and investigation of accidents;
> and such other matters concerned with the safety, regularity, and efficiency of air navigation as may from time to time appear appropriate.[75]

Therefore, the legal basis of the Annexes is Article 37 of the Chicago Convention.

---

74 Ann. 1 – Personnel Licensing, Ann. 2 – Rules of the Air, Ann. 3 – Meteorological Services, Ann. 4 – Aeronautical Charts, Ann. 5 – Units of Measurement, Ann. 6 – Operation of Aircraft, Ann. 7 – Aircraft Nationality and Registration Marks, Ann. 8 – Airworthiness of Aircraft, Ann. 9 – Facilitation, Ann. 10 – Aeronautical Telecommunications, Ann. 11 – Air Traffic Services, Ann. 12 – Search and Rescue, Ann. 13 – Aircraft Accident and Incident Investigation, Ann. 14 – Aerodromes, Ann. 15 – Aeronautical Information Services, Ann. 16 – Environmental Protection, Ann. 17 – Security, Ann. 18 – The Safe Transportation of Dangerous Goods by Air, Ann. 19 – Safety Management.

75 Chicago Convention, Art. 37.

The 19 Annexes are made up of Standards and Recommended Practices, which are commonly referred together as SARPs. There are currently over 12,000 SARPs found in the 19 Annexes that are managed by ICAO.[76] Annexes 2 and 7 contain only Standards, of which a definition can be found in Annex 17.

> Any specification for physical characteristics, configuration, materiel, performance, personnel or procedure, the uniform application of which is recognized as necessary for the safety or regularity of international air navigation and to which Contracting States will conform in accordance with the Convention; in the event of impossibility of compliance, notification to the Council is compulsory under Article 38 of the Convention.

Whereas, the 17 other Annexes contain a collection of Standards and Recommended Practices, with the latter also being defined in Annex 17.

> Any specification for physical characteristics, configuration, materiel, performance, personnel or procedure, the uniform application of which is recognized as desirable in the interests of safety, regularity or efficiency of international air navigation, and to which Contracting States will endeavour to conform in accordance with the Convention.

As the Annexes contain detailed technical rules on a wide range of areas, it is necessary to examine whether ICAO has developed SARPs that directly address and indirectly impact upon cybersecurity.

According to Annex 17, a definition does "not have an independent status but is an essential part of each Standard and Recommended Practice in which the term is used, since a change in the meaning of the term would affect the specification". Therefore, while they are included separately from the SARPs within each Annex, they are to be included as a part of them.

ICAO may also develop other documents.

1. Procedures for Air Navigation Services (PANS).[77]

---

76 ICAO, 'How ICAO Develops Standards', https://www.icao.int/about-icao/AirNavigationCommission/Pages/how-icao-develops-standards.aspx.

77 "Procedures for Air Navigation Services (or PANS) comprise operating practices and material too detailed for Standards or Recommended Practices – they often amplify the basic principles in the corresponding Standards and Recommended Practices." ICAO, 'Making an ICAO Standard', https://www.icao.int/safety/airnavigation/Pages/standard.aspx.

2. Regional Supplementary Procedures (SUPP).[78]
3. Guidance Material[79] (e.g. Manuals[80] and Circulars[81]).

These documents may also be relevant for this investigation into cybersecurity, so these will be looked at within Chapter III.4 where appropriate.

4.2 Annex 17 on Security

4.2.1 'Cyber Threats'

SAPRs on aviation security were adopted by the ICAO Council on 22 March 1974. These were assigned to Annex 17 titled 'Security' which "sets out the basis for the ICAO civil aviation security programme and seeks to safeguard civil aviation and its facilities against acts of unlawful interference".[82]

Annex 17 is kept under review by the Aviation Security Panel (AVSECP) – with the help of the Working Group on Annex 17, Threat and Risk Working Group, Working Group on Innovation in Aviation Security – and is currently in its tenth edition having been updated in response to the needs identified by States. The AVSECP is a group of experts as appointed by the Council and includes representatives from numerous States, as well as international organisations, such as the ACI, IATA, IFALPA and the International Criminal Police Organization. The panel, along with the ICAO Secretariat, develops ICAO "security policy and responses to emerging threats as well as strategies aimed at preventing future acts of unlawful interference".[83] Its work, however, has a strong focus on traditional security matters, i.e. physical security.

Annex 17 predominantly uses the term 'unlawful interference'. However, Amendment 12, which was added to the Annex in the 9th Edition that was published in March 2011,

---

78 "Regional Supplementary Procedures (or SUPPs) have application in the respective ICAO regions. Although the material in Regional Supplementary Procedures is similar to that in the Procedures for Air Navigation Services, SUPPs do not have the worldwide applicability of PANS".*Ibid.*

79 "Guidance Material is produced to supplement the SARPs and PANS and to facilitate their implementation. Guidance material is issued as Attachments to Annexes or in separate documents such [as] manuals, circulars and lists of designators/addresses. Usually it is approved at the same time as the related SARPS are adopted." *Ibid.*

80 "Manuals provide information to supplement and/or amplify the Standards and Recommended Practices and Procedures for Air Navigation Services. They are specifically designed to facilitate implementation and are amended periodically to ensure their contents reflect current practices and procedures." *Ibid.*

81 "Circulars make available specialized information of interest to Contracting States. Unlike manuals, circulars are not normally updated." *Ibid.*

82 ICAO, 'The Convention on International Civil Aviation Annexes 1 to 18', https://www.icao.int/safety/airnavigation/NationalityMarks/annexes_booklet_en.pdf.

83 ICAO, 'Security', https://www.icao.int/Security/Pages/default.aspx.

titled 'Proposals of the Committee on Unlawful Interference with the assistance of AVSECP and Council Action in Pursuance of Assembly Resolution A36-20', provided an addition to the terminology.

> This amendment includes provisions to further strengthen Standards and Recommended Practices in order to address new and emerging threats to civil aviation. The amendment includes the following: deployment of security equipment; security of air traffic service providers; training programmes and instructor certification system; random and unpredictable security measures; supply chain security; security for all-cargo operations; cyber threats; and definitions.[84]

While cyber threats to civil aviation were neither new nor emerging by 2011, Amendment 12 introduced the term 'cyber' to the Annexes for the first time.

The term 'cyber' appears only once in the substantive text of Annex 17. This is within Chapter 4 on Preventive Security Measures, more specifically in the title of 4.9 on Measures Relating to Cyber Threats. Here, two Recommendations were presented. However, the term 'cyber', which remains undefined, is not used in the substantive text of the two Recommendations. A revised text was then presented on 26 February to 16 March 2018 to the Council for adoption during its 213th Session. Amendment 16 to Annex 17 then became effective on 16 July 2018 and was applicable as of 16 November 2018. The new text will be examined below.

The first inclusion is Standard 4.9.1, which was originally a Recommended Practice, and states:

> Each Contracting State shall ensure that operators or entities as defined in the national civil aviation security programme or other relevant national documentation identify their critical information and communications technology systems and data used for civil aviation purposes and, in accordance with a risk assessment, develop and implement, as appropriate, measures to protect them from unlawful interference.[85]

84 Ann. 17 to the Chicago Convention, 2017, at xii.

85 The original text included in Ann. 17:

> Each Contracting State should, in accordance with the risk assessment carried out by its relevant national authorities, ensure that measures are developed in order to protect critical information and communications technology systems used for civil aviation purposes from interference that may jeopardize the safety of civil aviation.

Therefore, operators and other stakeholders which are identified at a national level must identify both their civil aviation:

1. critical information, and
2. communications technology systems and data,

and then develop and implement measures to protect them from unlawful interference. This, therefore, has some weaknesses:

1. There is a strong reliance on Contracting States to define the scope of the Standard, so fragmentation is a possibility among the different States.
2. It pertains only to measures. This is extremely vague as no further instructions are given. Thus, it could amount to, *inter alia*, the creation of new laws, the dissemination of information or training measures. This vagueness could result in fragmentation and potentially *de minimis* measures.
3. It is to protect critical information and communications technology systems used for civil aviation and, therefore, excludes non-critical information and technology systems. As a result, it does not cover all vulnerabilities found in the aviation context.
4. The measures must protect from only unlawful interference. Therefore, it leaves the attack methods that are not prohibited by law outside of its scope.
5. It has a security focus, so the safety aspect is not addressed.

As a result of these five limitations, the Standard has a limited impact on the regulation of cybersecurity. However, it is a vast improvement on the old text as it converted the Recommended Practice into a Standard, and it provides a broader scope.

Recommendation is 4.9.2 and states:

> Each Contracting State should ensure that the measures implemented protect, as appropriate, the confidentiality, integrity and availability of the identified critical systems and/or data. The measures should include, inter alia, security by design, supply chain security, network separation, and the protection and/or limitation of any remote access capabilities, as appropriate and in accordance with the risk assessment carried out by its relevant national authorities.[86]

---

86 Original text contained in Ann. 17:

> Each Contracting State should encourage entities involved with or responsible for the implementation of various aspects of the national civil aviation security programme to identify their critical information and communications technology systems, including threats and vulnerabilities thereto, and develop protective measures to include, *inter alia*, security by design, supply chain security, network separation, and remote access control, as appropriate.

The Recommendation then compliments the Standard as it gives guidance on how the implemented measures should protect. It is not surprising that 4.9.1. and 4.9.2. are limited in scope.

1. The binding status of SARPs is debateable and States can register non-compliance with ICAO.
2. They are the first direct step taken by ICAO to amend Annexes to address cybersecurity, so this should not mark the end.
3. They are contained in Annex 17 on Security so are unlikely to directly address safety.
4. The initial steps should prioritise areas that pose risk to life, such as critical information and communications technology systems, rather than secondary systems.

The two Recommendations mark the beginning, rather than the end. As they have since been amended, it can be expected that more amendments to the Annexes directly addressing cybersecurity will follow. This will be supported by the work of the ICAO Secretariat Study Group on Cybersecurity. Thus, it is a step in the right direction by this aviation stakeholder.

4.2.2 Aviation Security Manual

ICAO published Document 8973 Aviation Security Manual[87] to assist Member States "in implementing Annex 17 to the Chicago Convention by providing guidance on how to apply its Standards and Recommended Practices (SARPs)".[88] Annex 17 is periodically amended to address new threats and technological developments to reflect these changes.

Amendments to the Manual are focused on two areas. First, new guidance material is added to the text, so as to widen its scope to cover such areas as "unpredictability, behaviour detection techniques, landside security, and screening of persons other than passengers".[89] Second, existing guidance material has been updated to "better address the evolving threat to civil aviation, and includes updates on, *inter alia*, dangerous goods awareness training, screening technology and equipment, screening of vehicles and supplies, air cargo and mail secure supply chain measures, and threat and risk assessment methodology".[90]

Distribution of the Manual is highly restricted, so it is not possible to comment on its content. However, as Amendment 12 to Annex 17 specifically addresses cybersecurity,[91]

87 ICAO, 'Aviation Security Manual (Doc. 8973 – Restricted)', https://www.icao.int/Security/SFP/Pages/SecurityManual.aspx.
88 *Ibid.*
89 *Ibid.*
90 *Ibid.*
91 *See*, Chapter III.4.2.

it is likely that the Manual has also been amended to assist States in implementing cybersecurity. The importance of this document for cybersecurity and for the Member States that use it will continue to grow the more cybersecurity is explicitly added to Annex 17 and the more cybersecurity is linked to safety and security more generally.

4.2.3 Aviation Security Policy Section

Integral to the amendment of Annex 17 is the Aviation Security Policy (ASP) Section of the Aviation Security and Facilitation Office as it is tasked with the development of new and amended SARPs within the scope of security of international air transport. Among other things, the Aviation Security Policy Section manages its programmes by:

1. developing and updating Annex 17 to the Convention on International Civil Aviation (also known as the Chicago Convention);
2. managing and overseeing the work programme of the Aviation Security (AVSEC) Panel as well as its working groups;
3. updating and maintaining the ICAO Security Manual for Safeguarding Civil Aviation Against Acts of Unlawful Interference (Doc. 8973);

[...]

6 promoting implementation of security-related SARPs; and
7 providing support to various governing entities at ICAO, in particular the Assembly, Council and Committee on Unlawful Interference (UIC).
8 fostering cooperation on security matters with other United Nations agencies and international bodies, such as coordinating with the UN Counter Terrorism Committee (UN CTC), Counter-Terrorism Committee Executive Directorate (CTED), International Air Transport Association (IATA) and Airports Council International (ACI).[92]

Due to its role in developing, updating and promoting security-based SARPs, it clearly has an important role for cybersecurity as demonstrated above in this subchapter.

4.3 Unlawful Interference

As noted above, the term 'unlawful interference' is prominent in Annex 17, but it can also be found in other Annexes. No definition of 'unlawful interference' is provided in the Chicago Convention. However, this term is found in several of the Annexes and within the context of the Security Treaties as discussed below in Chapter III.7. Annex 17 provides a definition of the term 'unlawful interference' in Chapter 1.

92 ICAO, 'Aviation Security Policy Section', https://www.icao.int/Security/SFP/Pages/default.aspx.

Acts of unlawful interference. These are acts or attempted acts such as to jeopardize the safety of civil aviation, including but not limited to:
- unlawful seizure of aircraft,
- destruction of an aircraft in service,
- hostage-taking on board aircraft or on aerodromes,
- forcible intrusion on board an aircraft, at an airport or on the premises of an aeronautical facility,
- introduction on board an aircraft or at an airport of a weapon or hazardous device or material intended for criminal purposes,
- use of an aircraft in service for the purpose of causing death, serious bodily injury, or serious damage to property or the environment,
- communication of false information such as to jeopardize the safety of an aircraft in flight or on the ground, of passengers, crew, ground personnel or the general public, at an airport or on the premises of a civil aviation facility.

This covers a wide range of activities, which may be of relevance to cybersecurity.

From this definition, it is clear that the act needs to be unlawful, so the act must be prohibited by a source of law. For example, the Montreal Convention, pursuant to Article 1, requires States to makes certain acts an offence, including placing devices and communicating false information.[93] Whereas, the Tokyo Convention, on the other hand, refers to offences under national law, which may address cybersecurity in a way that has consequences for aviation.[94] Thus, it is possible to see how different cyberattacks may fall under the provided definition. However, as correctly recognised by ICAO, "not all cybersecurity issues affecting the safety of civil aviation are unlawful and/or intentional, and should, therefore, be addressed through the application of safety management systems".[95] Consequently, the criterion of unlawfulness provides a limitation.

Secondly, interference is undefined and harder to interpret within cybersecurity. A cyberattack that affects the victim physically, such as controlling the aircraft or results in the user being unable to use certain equipment, such as communication and navigation links, could be interpreted as interference. However, it is less clear whether eavesdropping, tapping, key logging and other non-intrusive passive attack methods constitute interference. Thus, this term has a serious weakness and does not suit the emerging threat of cyberattack.

93 *See*, Chapter III.7.2.3.

94 *See*, Chapter III.7.2.1.

95 ICAO, Working Paper, Assembly – 39th Session, Executive Committee, Addressing Cybersecurity in Civil Aviation, Presented by the Council of ICAO, A39-WP/17, EX/5, 30 May 2016.

Unlawful inference is not only contained in ICAO's security Annex, it is also found in the Safety Annexes. Annex 2 has been chosen as an example due to its importance in safeguarding safe international air transport and the persuasiveness of its Standards.[96] Annex 2 has a long history with the term 'unlawful interference' stating in 1973.

6th Edition
17. Council action in pursuance of Assembly Resolutions A17-10 and A18-10: Practices to be followed in the event that an aircraft is being subjected to unlawful interference. 17 December 1973, 17 April 1974 and 23 May 1974.

20. Air Navigation Commission: Time-keeping accuracy in ATS units and on board aircraft; use of SSR code 7500 in the event of unlawful interference. 17 April 1976, 17 August 1976 and 30 December 1976.

8th Edition
28. Air Navigation Commission: Definition of 'apron'; special procedures for use during unlawful interference. 16 March 1987, 27 July 1987 and 19 November 1987.

9th Edition
29. Visual Flight Rules Operations Panel, Third Meeting (1986); Secretariat; Visual Aids Panel, Eleventh Meeting (1987); Air Navigation Commission; amendments consequential to adoption of amendments to Annex 6: Operation of aircraft in mixed VFR/IFR environments; surface movement of aircraft and surface movement guidance and control; acts of unlawful interference; helicopters as intercepting aircraft. 12 March 1990, 30 July 1990 and 14 November 1991.

37. Separation and Airspace Safety Panel (SASP): Pilot procedures in the event of unlawful interference; editorial amendments. 28 February 2003.

10th Edition
40. Air Navigation Commission: Definitions and associated procedures for ADS-B, ADS-C and ADS-C agreement; pilot procedures in the event of unlawful interference. 26 February 2007 16 July 2007 and 22 November 2007.

A direct reference to the term is found in Section 3.7 titled 'Unlawful Interference' which contains two Standards.

96 Ann. 2 to the Chicago Convention, 2005, at 2.1.1.

> 3.7.1 An aircraft which is being subjected to unlawful interference shall endeavour to notify the appropriate ATS unit of this fact, any significant circumstances associated therewith and any deviation from the current flight plan necessitated by the circumstances, in order to enable the ATS unit to give priority to the aircraft and to minimize conflict with other aircraft.[97]

The obligation is, through the use of 'shall', to notify the ATS so they can take appropriate measures to ensure safety of other aircraft and the aircraft being subjected to the unlawful interference.

> 3.7.2 If an aircraft is subjected to unlawful interference, the pilot-in-command shall attempt to land as soon as practicable at the nearest suitable aerodrome or at a dedicated aerodrome assigned by the appropriate authority unless considerations aboard the aircraft dictate otherwise.[98]

This provides instruction to a pilot to land the aircraft. Here is a clear link between ICAO's 'security', via 'unlawful interference' and 'safety', which requires certain actions to be taken by the pilot to mitigate the risk emanating from the unlawful interference.

This is, however, not the only focus given by ICAO to unlawful interference. For example, Annex 6, Volume II on the 'Operation of Aircraft' covers the reporting of acts of unlawful interference.

> 2.9.2 Following an act of unlawful interference, the pilot-in-command shall submit a report of such an act to the designated local authority.[99]

---

97 Ann. 2 to the Chicago Convention, 2005, at 3-14.
Note 1. – Responsibility of ATS units in situations of unlawful interference is contained in Ann. 11.
Note 2. – Guidance material for use when unlawful interference occurs and the aircraft is unable to notify an ATS unit of this fact is contained in Attachment B to this Annex.
Note 3. – Action to be taken by SSR-, ADS-B- and ADS-C-equipped aircraft which are being subjected to unlawful interference is contained in Ann. 11, the PANS-ATM (Doc. 4444) and the PANS-OPS (Doc. 8168).
Note 4. – Action to be taken by CPDLC-equipped aircraft which are being subjected to unlawful interference is contained in Ann. 11, the PANS-ATM (Doc. 4444), and guidance material on the subject is contained in the Manual of Air Traffic Services Data Link Applications (Doc. 9694).

98 Ann. 2 to the Chicago Convention, 2005, at 3-14.
Note 1. – Requirements for State authorities with respect to aircraft on the ground that are subject to unlawful interference are contained in Ann. 17, Ch. 5, 5.2.4.
Note 2. – See 2.4 regarding the authority of the pilot-in-command of an aircraft.

99 Ann. 6 to the Chicago Convention, Operation of Aircraft, Vol. II – International General Aviation – Aeroplanes, Ninth Edition, July 2016, at 2.9-1.

Again, as with Annex 2, this example provides for notification to be given to a relevant stakeholder. However, this time, it does not prescribe actions to combat an on-going act of unlawful interference, rather it prescribes actions to be carried out after the act of unlawful interference. Standard 2.9.2, therefore, has a reactive dimension so the local authority can assess the situation and also a future proactive dimension as the local authority can use the information gained from the report to improve in the area.

Annex 12 on Search and Rescue includes 'unlawful interference' in a different way.

> 4.2.4 The search and rescue plans of operation shall contain details regarding actions to be taken by those persons engaged in search and rescue, including:
> [...]
> k) cooperative actions to be taken in conjunction with air traffic services units and other authorities concerned to assist aircraft known or believed to be subject to unlawful interference.[100]

Therefore, Standard 4.2.4 facilitates cooperation between relevant stakeholders within the context of search and rescue.

The last example, for the purpose of this subchapter, is taken from Annex 14 on Aerodromes.

> 3.14.1 An isolated aircraft parking position shall be designated or the aerodrome control tower shall be advised of an area or areas suitable for the parking of an aircraft which is known or believed to be the subject of unlawful interference, or which for other reasons needs isolation from normal aerodrome activities.[101]

Standard 3.14.1 refers to the handling of physical assets, notably the aircraft. Therefore, rather than prescribing certain actions of certain stakeholders, such as reporting and cooperation, it discusses the designation of space at an aerodrome and the positioning of aircraft known or believed to be the subject of unlawful interference.

It can be seen that, while the term 'unlawful interference' is defined, the Annexes take very different approaches to the handling this matter. As a result, the specifics of each use must be assessed, as a generalisation of the term's use cannot be made. Attention should

100 Ann. 12 to the Chicago Convention, Search and Rescue, Eighth Edition, July 2004, at 4.2.4.
101 Ann. 14 to the Chicago Convention, Aerodromes – Vol. I. Aerodrome Design and Operations, Eighth Edition, July 2018, at 3-29.

also be given as to whether the SARP is contained within a 'safety' or 'security' Annex as this will alter the context and implementation of the measures. However, what can be seen, this term could include cybersecurity under certain circumstances.

4.4 Electromagnetic Interference

Annex 8 on Airworthiness of Aircraft, which is linked to Article 31 of the Chicago Convention,[102] contains broad standards that set out "the minimum basis for the recognition by States of Certificates of Airworthiness for the purpose of flight of aircraft of other States into and over their territories, thereby achieving, among other things, protection of other aircraft, third parties and property".[103] Annex 8 requires special mention as, while it refers to 'interference', it does not adopt the more familiar ICAO terminology. For example, Part IIIB on Aeroplanes over 5,700 kg for which Application for Certification was Submitted on or after 2 March 2004 utilises the term 'electromagnetic interference'.[104]

> 6.1.2 The design of the instruments, equipment and systems required by 6.1.1 and their installation shall be such that:
> [...]
> c) electromagnetic interference between them is minimized.[105]

This is supported in Standard 6.5 on Electromagnetic Interference Protection.

> Aeroplane electronic systems, particularly flight-critical and flight-essential systems, shall be protected against electromagnetic interference from both internal and external sources.[106]

The Annex does not provide a definition of 'electromagnetic interference', however, it can be inferred from the wording of Standard 6.5 that this refers to internal interference, which would include such installed equipment, and external interference, such as television, radio and mobile phones. This is confirmed by Annex 10 on Aeronautical Telecommunications, which states "[a]ny transmission, emission, or reception of signs, signals,

102 Chicago Convention, Art. 31:
Every aircraft engaged in international navigation shall be provided with a certificate of airworthiness issued or rendered valid by the State in which it is registered.

103 ICAO, 'The Convention on International Civil Aviation Annexes 1 to 18', https://www.icao.int/safety/airnavigation/NationalityMarks/annexes_booklet_en.pdf.

104 *See*, for another potentially relevant Article, Chicago Convention, Art. 30.

105 Ann. 8 to the Chicago Convention, Airworthiness of Aircraft, Eleventh Edition, July 2010, at IIIB-6-1.

106 *Ibid.*, at IIIB-6-2.

writing, images and sounds or intelligence of any nature by wire, radio, optical or other electromagnetic systems".[107] As a result, if those elements that are considered electromagnetic cause interference in the course of a cyberattack, these Standards may be relevant for cybersecurity.[108]

It can be inferred that this is wider than unlawful interference as it should be read to include non-hostile interference. What is clear is that it must be interference from an 'electromagnetic' source. As interference, either unlawful or lawful, would have the same effect on the aircraft, it is argued here that such a distinction should not be made, rather the focus should be on the interference and mitigating it through airworthiness standards, and not whether there is intent.

It may be the case that those stakeholders responsible for the implementation of Annex 8 and 10 may not take measures to combat a cyberattack, as such an act is traditionally dealt with by ICAO in Annex 17. Therefore, despite the possible relevance of the 'safety' Annexes for cybersecurity, implementation may prevent cybersecurity being applied due to the traditional dichotomy in aviation regulation.

4.5 *Filling the Gaps*

The term 'cyber threats' directly relates to cybersecurity, and a study of the two similar terms – 'unlawful interference' and 'electromagnetic interference' – show that they are also relevant. It is, therefore, possible to read cybersecurity into SARPs even where they were not drafted to specifically tackle such an issue. This conclusion is not limited to these two terms as, for example, Annex 10 on Aeronautical Telecommunications utilises several words, which could be potentially relevant.[109] The terms 'authorised', 'accuracy',[110] 'integrity',[111] 'availability' and 'functionality'[112] also feature in Annex 10 and

107 Ann. 10 to the Chicago Convention, Aeronautical Telecommunications, Vol. II – Communication Procedures Including those with PANS Status, Seventh Edition, July 2016, at 1-4.

108 *See*, J. Sun, J. Liu, Y. Yang, S. Fan and W. Yu, *China Satellite Navigation Conference (CSNC) 2017 Proceedings*, Vol. 1, (Springer, 2017), at 646.

109 *See*, by way of example of the term 'authorised', Ann. 10 to the Chicago Convention, Aeronautical Telecommunications, Vol. III – Communication Systems, Second Edition, July 2007.

110 *See, e.g.*, Ann. 15: "Accuracy. A degree of conformance between the estimated or measured value and the true value." Ann. 15 to the Chicago Convention, Aeronautical Information Services, Fifteenth Edition, July 2016, at 1-1.

111 *See, e.g.*, Ann. 15: "Integrity (aeronautical data). A degree of assurance that an aeronautical data and its value has not been lost or altered since the data origination or authorized amendment." *Ibid.*, at 1-5.

112 *See, e.g.*, Ann. 4: "Aerodrome Terrain and Obstacle Chart – ICAO (Electronic) – Function, Availability, Identification, Chart coverage, Chart content, Accuracy and resolution, Electronic functionality, Chart data product specifications". Ann. 4 to the Chicago Convention, Aeronautical Charts, Eleventh Edition, July (2009), at 5-1.

these are well established in cybersecurity vocabulary.[113] Annex 9 also uses terms, such as in the context of security travel documents "unlawfully altered, replicated or issued".[114] The current reach of the Annexes in terms of cybersecurity is, thus, much further than that which *prima facia* presents itself.

The broad scope of the Annexes, varied language, and goal of promoting safety and security should not be overlooked. In other words, cybersecurity discussions should not only focus on the applicability of Annex 17 nor should regulatory efforts by ICAO to address cybersecurity be focused on this one Annex alone. Cybersecurity and cyberattacks are complex and multifaceted. As a result, each SARP should be individually assessed to establish whether it applies to cybersecurity within its specific context. Whether this will be done by ICAO, however, remains to be seen.

4.6 Limited Scope of Annexes

It has been shown that the Annexes are relevant for cybersecurity and it is argued that the Annexes will continue to be a tool to promote cybersecurity and mitigate cyberattacks. This is likely to take place both with explicit reference to cybersecurity, notably that in Annex 17, and with indirect reference, such as that in the other Annexes. However, the impact that SARPs can have is limited; not only in the wording, such as that already highlighted in Chapter III.4, but also generally.

First, pursuant to Article 38 of the Chicago Convention, a Member State may depart from a Standard provided that it follows the notification procedure specified in that Article. However, "[s]tates do not always comply with this obligation: they may depart from, that is, not implement and enforce a Standard without notifying ICAO of such a move".[115] ICAO conducts audits through the Universal Safety Oversight Audit Programme, in order to "determine Member States' safety capabilities and the status of States' implementation of all safety-relevant ICAO Standards and Recommended Practices (found in 17 of the 19 ICAO Annexes), associated procedures, guidance material, and best safety prac-

113 *See*, as way of example, National Initiative for Cybersecurity Careers and Studies, 'Glossary', https://niccs.us-cert.gov/glossary. *See, e.g.*, Ann. 4:"Area navigation based on performance requirements for aircraft operating along an ATS route, on an instrument approach procedure or in a designated airspace. Note. – Performance requirements are expressed in navigation specifications (RNAV specification, RNP specification) in terms of accuracy, integrity, continuity, availability and functionality needed for the proposed operation in the context of a particular airspace concept." Ann. 4 to the Chicago Convention, 2009, at 1-7.

114 Ann. 9 to the Chicago Convention, Facilitation, Fifteenth Edition, October (2017), at 3.7.

115 P. Mendes de Leon, 2017, at 24.

tices".[116] There is also the Universal Security Audit Programme Continuous Monitoring Approach, which focuses on Annex 17 and the security-related provisions of Annex 9 – Facilitation.[117] These checks show that there is not 100 percent adherence to Standards nor is ICAO notified of every departure by States. This highlights that there is not universal application of SARPs by States, with departures being either notified to ICAO or being unreported, so Article 38 limits the effectiveness of SARPs and thus their impact on cybersecurity.

Second, as SARPs are not found in the text of the Chicago Convention and are instead developed as a consequence of it; they do not have treaty status. This has resulted in diverging opinions with regard to the legal status of SARPs in courts,[118] in literature[119] and in practice. It is not within the scope of this work to discuss this issue; however, it is important to highlight as it shows another limitation to the Annexes, and their ability to promote cybersecurity and mitigate the impact of cyberattacks as their binding force is questionable.

Third, the existing SARPs applicable to cybersecurity are fragmented and do not cover every aspect. ICAO can develop new SARPs to fill the gaps, but this will take time as "it takes approximately two years for an initial proposal for a new or improved Standard, Recommended Practice or procedure to be formally adopted or approved for inclusion in an Annex or a PANS".[120] The time frame is, however, flexible and can be "expanded or compressed depending on the nature and priority of the proposal under consideration".[121] If cybersecurity-based SARPs take a reactive approach, then an average two-year development phase is simply too long for the fast evolving and growing area of cybersecurity. The two-year development time also must be supplemented with the pre-proposal stage, whereby ICAO becomes aware of an issue and then takes steps to bring a change to the Annexes. As a result, even if States adopt the SARPs and strictly adhere to them, it will be some time before they fully cover the existing threat landscape and are likely to become outdated due to the fast-evolving topic.

---

116 ICAO, 'Frequently Asked Questions About USOAP', https://www.icao.int/safety/CMAForum/Pages/FAQ.aspx; and ICAO, Universal Safety Oversight Audit Programme Continuous Monitoring Manual, Doc. 9735 AN/960 (2011). *See*, J. Huang, 2009, at 69-71.

117 ICAO, 'The Universal Security Audit Programme Continuous Monitoring Approach (USAP-CMA) and its Objective', https://www.icao.int/Security/USAP/Pages/default.aspx. *See*, J. Huang, 2009, at 71-72.

118 *See*, P. Mendes de Leon, 2017, at 25.

119 *See*, J. Huang, 2009, at 58-68.

120 ICAO, 'How ICAO Develops Standards', https://www.icao.int/about-icao/AirNavigationCommission/Pages/how-icao-develops-standards.aspx.

121 *Ibid.*

Finally, the source of cyber-relevant SARPS is fragmented between the two regimes of 'safety' and 'security'. This division is encapsulated in Annex 2 whereby it states "technical difficulties or unlawful interference".[122] The prior referring to safety and the latter to security. Thus, it may be the case that 'safety' SARPs may be implemented without cybersecurity being considered, and for 'security' SARPs to ignore safety concerns. This could then produce vulnerabilities and should be addressed by ICAO to make stakeholders aware.

5 International Civil Aviation Organization

5.1 *Introductory Remarks*

ICAO is a UN specialised agency, established by the Chicago Convention and headquartered in Montreal, Canada, tasked to manage the administration and governance of the Chicago Convention. ICAO works with the Chicago Convention's 193 State Parties and other stakeholders in order to fulfil its aims and objectives set out in Article 44 the Chicago Convention.

a. Insure the safe and orderly growth of international civil aviation throughout the world;
b. Encourage the arts of aircraft design and operation for peaceful purposes;
c. Encourage the development of airways, airports, and air navigation facilities for international civil aviation;
d. Meet the needs of the peoples of the world for safe, regular, efficient and economical air transport;
e. Prevent economic waste caused by unreasonable competition;
f. Insure that the rights of contracting States are fully respected and that every contracting State has a fair opportunity to operate international airlines;
g. Avoid discrimination between contracting States;
h. Promote safety of flight in international air navigation;
i. Promote generally the development of all aspects of international civil aeronautics.[123]

In addition to SARPs, as mentioned above in Chapter III.4, ICAO has also been active in cybersecurity in other areas.[124] This subchapter will examine these initiatives.

122 Ann. 2 to the Chicago Convention, 2005, at ATT A-1.
123 *See*, ICAO, 'About ICAO', https://www.icao.int/about-icao/Pages/default.aspx.
124 *See*, for a general descriptive overview, R. Abeyratne, 'Aviation Cyber Security: A Constructive Look at the Work of ICAO', 41 *Air and Space Law* 25-40 (2016).

5.2 *General Assembly Resolutions*

The ICAO General Assembly is the sovereign body of the organisation and meets, at least once every three years, to discuss and establish its worldwide policy for the upcoming triennium. The 193 State Parties to the Chicago Convention, as well as a number of international organisations, are invited to participate at the Assembly. During its meetings, known as 'Sessions', the "ICAO's complete work programme in the technical, economic, legal and technical cooperation fields is reviewed in detail".[125] During this time, Resolutions are adopted and Working Papers can be presented by States or international organisation on topics specified on the Agenda.

The ICAO 39th Assembly took place at the end of 2016 and was attended by over 2,000 delegates that represented 185 Members States, as well as observer delegations from international, regional and other organisations.[126] In total, 38 Resolutions were produced by the Assembly. Several of these may be indirectly relevant for cybersecurity, such as A39-10 on the Promotion of the Beijing Convention and the Beijing Protocol of 2010. While the Beijing Convention and the Beijing Protocol have relevance for cybersecurity, as discussed in Chapter III.7, the promotion of these instruments found in Resolution A39-10 is only an indirect link. In cases where an indirect link is present between cybersecurity and a Resolution, these will not be discussed as ICAO's general safety and security system, and its link to cybersecurity is well discussed in other parts of Chapter III. Despite potential indirect connections, the ICAO 39th Assembly introduced the first Assembly Resolution focused on cybersecurity.

5.2.1 First Ever Cybersecurity Resolution

Working Paper A39-WP/17 EX/5: Addressing Cybersecurity in Civil Aviation, which was presented by the Council of ICAO, contains a Draft Assembly Resolution Addressing Cybersecurity in Civil Aviation in an effort to further promote a consistent and coherent approach in managing cyber threats, and risks on how to better address cybersecurity in civil aviation. The result was the October 2016 Resolution A39-19: Addressing Cybersecurity in Civil Aviation.

> The Resolution aims to address cybersecurity through a horizontal, crosscutting and functional approach, reaffirming the importance and urgency of pro-

125 ICAO, 'About the Assembly', https://www.icao.int/Meetings/a39/Pages/default.aspx.

126 ICAO, Fourth African-Indian Ocean (AFI) Week Second Security and Facilitation Symposium, Session 3: 39th ICAO Assembly Resolutions Related to Aviation Security and Facilitation, 2nd AFI SECFAL Symposium 2017, Wednesday, 24 May 2017, Gaborone, Botswana, 19 May 2017.

> tecting civil aviation's critical infrastructure systems and data against cyber threats.[127]

Further to this, it "also seeks to obtain global commitment to action by ICAO, its Member States and industry stakeholders"[128] to work to collaboratively and systematically address cybersecurity in civil aviation and mitigate the associated threats and risks. The document is extremely important in gauging ICAO's understanding of cybersecurity and how it plans to deal with it for the next three years and beyond.

In the Preamble to the Resolution, ICAO picks up on some important issues. Firstly, ICAO clearly sees the need for and benefit of collaborating with other stakeholders in order to develop an effective and coordinated global framework to address the challenges of cybersecurity, which it links back to the Industry High Level Group (IHLG).

Secondly, ICAO is conscious that the factual situation is important to understand in order to properly promote cybersecurity and address cyberattacks in a prompt way. For example, ICAO notes "that [the] aviation sector is increasingly reliant on the availability of information and communications technology systems, as well as on the integrity and confidentiality of data"[129] and recognises "the multi-faceted and multi-disciplinary nature of cybersecurity challenges and solutions".[130] Further, ICAO is mindful that the risk is constantly evolving with different attack methods being developed, for different motives. Finally, it also appreciates not only the complexity of cybersecurity and cyberattacks, but also the complexity of the 'global aviation system'. This is extremely important, as outlined in Chapter I.

Third, it shows that ICAO regards the issue as a high priority as it reaffirms "the importance and urgency of protecting civil aviation's critical infrastructure systems and data against cyber threats".

Finally, and very importantly, ICAO recognises "that not all cybersecurity issues affecting the safety of civil aviation are unlawful and/or intentional, and should, therefore, be addressed through the application of safety management systems".[131] Therefore, ICAO directly references the need to address the safety side of cybersecurity.

---

127 ICAO, Journal, Global Aviation Security Plan (GASeP) to be Developed, Vol. 71(4), at 16.
128 *Ibid.*
129 ICAO, Assembly Resolutions in Force (as of 6 October 2016), Doc. 10075, (2017).
130 *Ibid.*
131 *Ibid.*

The Resolution is split into two directions. Firstly, the Assembly looks externally and requests its Member States and industry stakeholders to take certain steps.

a. Identify the threats and risks from possible cyber incidents on civil aviation operations and critical systems, and the serious consequences that can arise from such incidents;
b. Define the responsibilities of national agencies and industry stakeholders with regard to cybersecurity in civil aviation;
c. Encourage the development of a common understanding among Member States of cyber threats and risks, and of common criteria to determine the criticality of the assets and systems that need to be protected;
d. Encourage government/industry coordination with regard to aviation cybersecurity strategies, policies, and plans, as well as sharing of information to help identify critical vulnerabilities that need to be addressed;
e. Develop and participate in government/industry partnerships and mechanisms, nationally and internationally, for the systematic sharing of information on cyber threats, incidents, trends and mitigation efforts;
f. Based on a common understanding of cyber threats and risks, adopt a flexible, risk-based approach to protecting critical aviation systems through the implementation of cybersecurity management systems;
g. Encourage a robust all-round cybersecurity culture within national agencies and across the aviation sector;
h. Determine legal consequences for activities that compromise aviation safety by exploiting cyber vulnerabilities;
i. Promote the development and implementation of international standards, strategies and best practices on the protection of critical information and communications technology systems used for civil aviation purposes from interference that may jeopardize the safety of civil aviation;
j. Establish policies and allocate resources when needed to ensure that, for critical aviation systems: system architectures are secure by design; systems are resilient; methods for data transfer are secured, ensuring integrity and confidentiality of data; system monitoring, and incident detection and reporting, methods are implemented; and forensic analysis of cyber incidents is carried out; and
k. Collaborate in the development of ICAO's cybersecurity framework according to a horizontal, cross-cutting and functional approach involving air navigation, communication, surveillance, aircraft operations and airworthiness and other relevant disciplines.

There is a clear focus on cooperation between different stakeholders; the sharing of information; identification of risks; and encouraging a cybersecurity culture through *inter alia*, designated responsibilities, promotion of standards and the allocation of resources. It also covers the relationship between ICAO and States, however, more could be said on assisting States with the development of national or regional cybersecurity rules, which, as shown in Chapter II, is often a missing piece in cybersecurity, and on promoting State compliance with cyber-based SARPs.

The second step then looks internally at ICAO as it compels the Secretary General to take certain action.

a. Assist and facilitate States and industry in taking these actions; and
b. Ensure that cybersecurity matters are fully considered and coordinated across all relevant disciplines within ICAO.

The role of ICAO in assisting States is vital and this is correctly raised. Point (a) is limited to providing help to stakeholders, however, point (b) extends the scope of ICAO's role under Article 44 of the Chicago Convention as it touches on the full competencies of ICAO and, thus, may include amending existing or creating new Annexes.

5.2.2 Working Papers

Resolution A38-15 on Consolidated Statement of Continuing ICAO Policies Related to Aviation Security, as Presented by the Council of ICAO, was adopted in 2013 by the 38th Session of the ICAO Assembly. Pursuant to Clause 2 of the Resolution, the consolidated Statement is to be reviewed at each Ordinary Session of the Assembly. Working Paper A39-WP/16 EX/4 presents a revision and includes the term 'cyber' twice.

> Whereas a clear need exists for the strengthening of security to be applied to all phases and processes associated with the carriage of persons, their cabin and hold baggage, cargo, mail, courier and express parcels, and in protecting civil aviation against cyber-attacks and cyber threats.[132]

This correctly highlights the need to include cybersecurity within security. However, it is indicative of ICAO's treatment of cybersecurity as a security matter but not as a safety matter.

132 ICAO, Working Paper, Assembly – 39th Session, Executive Committee, Consolidated Statement of Continuing ICAO Policies Related to Security, A39-WP/16, EX/4, 3 June 2016, at A-6.

A39-WP/99 on Cyber Resilience in Civil Aviation was presented by the US and Slovakia on behalf of the EU, ECAC Member States and Eurocontrol.[133] The Working Paper focuses on "the importance of defining an appropriate cybersecurity approach in civil aviation", but recognises that "additional efforts are still required to increase global awareness and to further develop globally coherent cyber resilience approaches for the aviation system".[134] To achieve this, the Working Paper invites the Assembly to take five actions.

1. Address cyber resilience in civil aviation in a comprehensive manner.
2. Promote awareness of cyber threats and vulnerabilities.
3. Facilitate information sharing.
4. Consider what the required steps are for the production of guiding principles for managing current and future cyber-based threats and vulnerabilities.
5. Instruct existing panels and expert groups, which could include the AVSEC Panel,[135] to take the above into account as necessary.

A39-WP/175 on Civil Aviation Cyber-Security: Possible Actions by Regulators and Stakeholders, which was presented by Singapore, on behalf of 23 States,[136] complements A39-WP/99.[137] The Working Paper acknowledges the important work already undergone by stakeholders, including ICAO, but notes that many stakeholders are still struggling with cybersecurity matters. As a result, it invites the Assembly to note the contents of the Working Paper, including the nine proposed stakeholder actions; urges ICAO to "establish a global framework for cyber-security for civil aviation stakeholders"; and urges ICAO to "identify a body within the ICAO to work on the global framework and for coordination with other relevant ICAO bodies and other stakeholders". The body has not yet been identified, but it could involve the AVSEC Panel or the Threat and Risk Working Group.[138]

Working Paper A39-WP/187 on Cyber Security Defense Strategy was presented by Civil Air Navigation Services Organization (CANSO) and "underlined the importance of the protection of all digital information and systems within aviation networks, called for

133 ICAO, Working Paper, Assembly – 39th Session, Executive Committee, Technical Commission, Cyber Resilience in Civil Aviation, A39-WP/99 EX/45, TE/26, Revision 1, 26 July 2016.

134 *Ibid.*, at 1.

135 *See*, ICAO, 'AVSEC Panel', https://www.icao.int/Security/Pages/avsec-panel.aspx.

136 Argentina, Belgium, France, Guyana, Lao People's Democratic Republic, Namibia, Nauru, Nepal, Netherlands, Nigeria, Republic of Moldova, Russian Federation, Saint Lucia, Saudi Arabia, Senegal, Sierra Leone, Singapore, South Africa, Switzerland, The former Yugoslav Republic of Macedonia, Trinidad and Tobago, UAE and UK.

137 ICAO, Working Paper, Assembly – 39th Session, Executive Committee, Civil Aviation Cyber-Security: Possible Actions by Regulators and Stakeholders, A39-WP/175, EX/64, 8 August 2016.

138 *See*, ICAO, Working Paper, Assembly – 39th Session, Executive Committee, 'Cyber Security Defense Strategy', A39-WP/1871, EX/72, 10 August 2016, at 2.

greater coordination among the many aviation communities using ICT systems, and recognised that the protection of relevant assets against cyberattacks may be carried out differently across the aviation system, in accordance with stakeholders' priorities".[139] CANSO invited the Assembly to agree on the following recommendations contained in the Working Paper:

a. consider the content of the paper and endorses it, recognizing the relevance and the possible impact to civil aviation;
b. agree that ICAO develop a new strategy on aviation cybersecurity as a vertical domain within aviation security and issue guidance material, aimed at harmonization of the current initiatives both at regional and national level;
c. recommend Member States to focus their attention on vulnerabilities in systems and networks rather than on threats and to consider the implications on the entire civil aviation industry;
d. recommend that a forum is set up for aviation firms to share best practices in a secure/trusted environment so that effective tools and techniques that enhance security in the current and future technology estates can be shared;
e. recommends Member States to review recent initiatives presented during AVSECP/27 in order to elevate to a Standard the current Recommended Practices 4.9.1 and 4.9.2 of Annex 17; and
f. take note of the CANSO 'Cyber Security and Risk Assessment Guide'.

CANSO greatly stresses that both ICAO and States are important as a full legal approach is required for enhancing cybersecurity and mitigating the effects of cyberattacks.

A39-WP/236 on Coordinating Cybersecurity Work was presented by the International Coordinating Council of Aerospace Industries Associations (ICCAIA). In the Working Paper, the ICCAIA noted the work already carried out in the area of cybersecurity, in which it has participated in parts, and invited the Assembly to "request the Council to have the Secretariat form an expert group, as soon as practicable, to assist and advise ICAO in steering and coordinating the work on cyber security, cyber safety and cyber resilience".[140]

139 ICAO, Working Paper, Assembly – 39th Session, Executive Committee, 'Cyber Security Defense Strategy', A39-WP/1871, EX/72, 10 August 2016.
140 ICAO, Working Paper, Assembly – 39th Session, Executive Committee, 'Coordinating Cybersecurity Work', A39-WP/236, TE/91, 26 August 2016, Revision No. 1, 20 September 2016.

In addition, there are three Working Papers, Information Papers that cover cybersecurity within the titles.

1. A39-WP/296: Surveillance of Remotely Piloted Aircraft Systems (RPAS) and Cybersecurity.[141]
2. A39-WP/337: Aircraft Certification Cybersecurity Regulatory Efforts.[142]
3. A39-WP/370: A Cybersecurity Architectural Approach for Legacy- and SWIM-Based CNS/ATM Systems.[143]

These provide detailed information on a specific topic as understood by their respective presenter.

In addition to the explicit references to cybersecurity, several of the Working Papers may be interpreted to be relevant.

1. A39-WP/145 EX/54: Development and Updating of Standards and Guidance Material on New and Emerging Threats.
2. A39-WP/180 EX/67: Global Strategies for Addressing Insider Risk.
3. A39-WP/198 EX/74: Landside Security Policies.
4. A39-WP/310 EX/113: Security Measures for the Landside Areas of Airports.
5. A39-WP/98 EX/43: Priorities for Aviation Security.

A39-WP/145 EX/54, for example, encourages ICAO to focus on new threats. While cybersecurity is not included, this could potentially be inferred. Second, A39-WP/198 EX/74 tackles landside security which could have relevance, such as reducing threats from insiders as demonstrated in Chapter II.10. Finally, another indirect way is via A39-WP/98 EX/43, which encourages States to implement Annex 17. This demonstrates that, while cybersecurity is not directly referenced in the paper, it may still be indirectly relevant as Annex 19 includes 'cyber threats'.

The 40th ICAO General Assembly will be held between 24 September and 4 October 2019. While cybersecurity does not feature on its Provisional Agenda for the 40th Session of the ICAO Assembly, it will certainly be a topic raised by invited stakeholders. As a

141 ICAO, Working Paper, Assembly – 39th Session, Technical Commission, 'Surveillance of Remotely Piloted Aircraft Systems (RPAS) and Cybersecurity', Information Paper, A39-WP/2961, TE/125, 26 August 2016, Revision No. 1, 8 September 2016.

142 ICAO, Working Paper, Assembly – 39th Session, Executive Committee and Technical Commission, 'Aircraft Certification Cybersecurity Regulatory Efforts', Information Paper, A39-WP/337, EX/130, TE/145, 2 September 2016.

143 ICAO, Working Paper, Assembly – 39th Session, Technical Commission, 'A Cybersecurity Architectural Approach for Legacy- and SWIM-Based CNS/ATM Systems', Information Paper, A39-WP/370, TE/159, 8 September 2016.

result, ICAO's work in this area is likely to continue and increase as a result of the next General Assembly meeting.

5.3 *Global Aviation Security Plan*

The ICAO Comprehensive Aviation Security Strategy (ICASS) was approved by the Council during its 189th Session, and endorsed by the 37th Session of the Assembly.[144] Working Paper A39-WP 14 highlights the outcomes and achievements of ICASS's implementation during the 2014-2016 triennium.[145] ICASS has since been replaced by The Global Aviation Security Plan (GASeP). The GASeP:

> provides the foundation for States, industry, stakeholders and ICAO to work together with the shared and common goal of enhancing aviation security worldwide and achieving five key priority outcomes, namely: a) enhance risk awareness and response; b) develop security culture and human capability; c) improve technological resources and innovation; d) improve oversight and quality assurance; and e) increase cooperation and support.[146]

In doing so, the overall aim is to raise the level of implementation of Annex 17 through actions at a national, regional and global level.

The GASeP contains an extensive Roadmap in Appendix A which contains specific measures and tasks. Specific to cybersecurity are 1.A and 1.B which take a holistic approach to cybersecurity.

---

144 ICAO, Working Paper, Assembly – 37th Session, Executive Committee, 'ICAO Comprehensive Aviation Security Strategy (ICASS) Strategic Plan of Action 2011-2016', A37-WP/18 EX/2 28 June 2010.

145 ICAO, Working Paper, Assembly – 39th Session, Executive Committee, 'ICAO Comprehensive Aviation Security Strategy (ICASS)', A39-WP 14, EX/2, 30 May 2016.

146 ICAO, Global Aviation Security Plan, November 2017, at iii.

1.A Identify and address cybersecurity threats to civil aviation's critical infrastructure, data and information and communication technology systems through collaboration using horizontal, cross-cutting and functional approaches to achieve an acceptable and commensurate cyber resilience capability on a global level. It should involve air navigation, communication, surveillance, aircraft operations and airworthiness and other relevant disciplines to ensure the safety and security of civil aviation operations in full alignment with ICAO's Global Air Navigation Plan (GANP) and Global Aviation Safety Plan (GASP).

1.B Monitor and address emerging and evolving risks, such as cyber security, remotely piloted aircraft systems (RPAS), and risks arising from conflict zones.

Therefore, the measures imposed here cover identification of current and emerging threats, and take Action to combat cyber threats.

The Road Map addresses aviation security and as a result, other tasks and actions may be applicable, such as 1.Y which proposes to "develop a Manual on Investigating Acts of Unlawful Interference Against Civil Aviation". Therefore, the scope of the Road Map to cybersecurity is potentially wide and must be considered.

5.4 *European Aviation Security Training Institute*

Due to the increased importance that security has developed in international air transport, ICAO and ECAC committed to implementing the highest standards in civil aviation security, established a training centre in the European region for the enhancement of aviation security. This training centre has been called the European Aviation Security Training Institute (EASTI) and its objectives are to:

1. conduct aviation security classroom and e-learning training courses covering a wide range of issues. This scope is being enlarged to safety and facilitation issues.
2. build a European Information and Documentation Centre on Aviation Security, i.e. a publications library, a newspaper archive and a video library. Today it already provides the largest and most complete book, video and DVD library on aviation security worldwide.
3. operate as a research centre for training methods and equipment. Currently EASTI operates a CBT-system for selection, training and certification for operators of detection equipment, and other security personnel.

4. function as a European e-learning centre on aviation security.
5. cooperate closely with other ICAO ASTC's and support national (non commercial) AVSEC training centres.
6. cooperate closely with the stakeholders.
7. contribute to ICAO, ECAC and EC projects as regards the development of training courses and training aids. Special attention is given to the ICAO Standardised Training Packages (STP's) of which EASTI develops the European versions by incorporating ECAC recommendations and EU regulations.
8. coordinate aviation security training efforts in the European Region.
9. promote a joint partnership between all AVSEC training centres, host discussion forums and organise workshops between AVSEC training centres on training methodologies.
10. promote interregional cooperation in the field of training by organising joint courses and workshops and by setting up a worldwide e-learning network.[147]

The classroom and e-learning courses cover a wide range of aviation security topics. While none specifically address cybersecurity, this can be expected to change with the growing attention that ICAO and ECAC afford the topic.

5.5 *External Support*

Attempts to regulate aviation pre-date the Chicago Convention with key examples including the Paris Convention 1919, and the regional Ibero-American Convention on Air Navigation (Madrid Convention 1926)[148] and the Pan American Convention on Commercial Aviation (Havana Convention 1928).[149] These were superseded by the Chicago Convention.[150] While the Chicago Convention only marginally touches upon cooperation outside of ICAO, Article 37 of the Convention on the adoption of international standards and procedures states:

147 EASTI, 'Objectives', https://easti.eu/objectives/.

148 Ultimately, this Convention had a limited impact because of the restricted number of ratifications that it received; whereas 21 States (Spain, Portugal and 19 Latin American countries) attended the Madrid Conference, only 7 States (Argentina, Costa Rica, Dominican Republic, El Salvador, Mexico, Spain and Paraguay) deposited their instruments of ratification. The Madrid Convention 1926 never came into force.

149 American Convention on Commercial Aviation 1928. The Convention was signed by twenty-one States and was finally ratified by sixteen: Bolivia, Brazil, Chile, Costa Rica, Cuba, the Dominican Republic, Ecuador, Guatemala, Haiti, Honduras, Mexico, Nicaragua, Panama, Uruguay, US, and Venezuela.

150 Chicago Convention, Art. 80.

> Each contracting State undertakes to collaborate in securing the highest practicable degree of uniformity in regulations, standards, procedures, and organization in relation to aircraft, personnel, airways and auxiliary services in all matters in which such uniformity will facilitate and improve air navigation.

A legal basis can, therefore, be implicitly found in Article 37 for States to collaborate with other States to ensure high standards of aviation safety by way of uniformity, whether this be bilaterally, through regional cooperation, fragmented global cooperation or worldwide cooperation. It is for the concerned States to "determine the legal form and institutional structure that best fits the needs and characteristics of their specific region".[151] As a result, numerous international and regional organisations have been created to promote different aspects of aviation, such as safety and security.

There are several aviation organisations across the world that aim to promote aviation safety and security. For example, the Regional Initiative Projects South Asia Regional Initiative (SARI):[152]

> was created to update insufficient and outdated legislation, regulations and guidance on aviation safety and provide a legal foundation for regional cooperation in South Asia. It is also aimed at assisting airlines and regulators in interpreting rules. One of the main goals of SARI is to ensure that the region retains qualified technical personnel and to eliminate duplication of efforts and confusion of functions within civil aviation authorities.[153]

However, SARI's scope of work is currently limited to airworthiness matters and is growing with regard to flight operations rules. Under this scope, there is a clear link with cybersecurity. However, while aware of this issue, it has no short or medium terms plans concerning cybersecurity and this is unlikely to change anytime soon.

---

151 ICAO Safety Oversight Manual; Doc. 9734-AN/959, at Part B, Forward.

152 Bangladesh, Bhutan, India, Maldives, Nepal, Pakistan and Sri Lanka. SARI, 'SARI Project', http://www.searif-sari.org/SARI-SEARIF/SARI.html. *See*, South East Asia Regional Initiative Forum (SEARIF), http://www.searif-sari.org/SARI-SEARIF/SEARIF.htm. This includes: Brunei Darussalam, Cambodia, Indonesia, Lao PDR, Malaysia, the Philippines, Singapore, Thailand and Vietnam.

153 R. Abeyratne, *Strategic Issues in Air Transport: Legal, Economic and Technical Aspects*, First Edition, (Springer, 2012), at 151.

The case of SARI is not unique as, for example, the Pacific Aviation Safety Office (PASO),[154] which was established on 11 June 2005 by the Pacific Islands Civil Aviation Safety and Security Treaty is at the same stage. The organisation is based in Port Vila, Vanuatu and is managed by the General Manager, who oversees the running of the office, and three aviation inspectors specialised in airworthiness, aviation security and flight operations. The aviation inspectors provide technical assistance to the State Parties within the respective areas of airworthiness, flight operations, airports, security and personnel licencing. Furthermore, PASO is also involved in air navigation and accident investigation. Again, this regional organisation has not taken on cybersecurity, with this unlikely to change anytime soon, so it is for the States themselves to deal with cybersecurity obligations, despite the relevance of the topic for its area of competencies.

There is a growing tendency among States to mandate regional organisations with safety and security-related tasks. These pertain to operation, regulation, enforcement and supervision. Cybersecurity is an emerging area within aviation regulation, so some organisations have yet to begin discussing the topic.

The general treatment of cybersecurity by such organisations under the domain of security has resulted in the topic not being discussed in many regional organisations as their scope is often on core aviation safety aspects, such as airworthiness, maintenance and licencing. However, as the rules developed by EASA have shown, cybersecurity is relevant for these areas and should then be taken up by these organisations.

Annex 17 contains a direct reference to cybersecurity, so the regional organisations should address this within their SARP-based activities. Therefore, it can be reasonably expected that the role of regional organisations will increase in importance. However, this may be too slow to have an efficient and effective impact on cyber threats.

5.6 *Secretariat Study Group on Cybersecurity*

The 39th Session of the ICAO Assembly, due to the increasing threat of cyberattacks, "called for a coordinated approach to achieve an acceptable and commensurate cyber resilience capability on a global scale".[155] Resulting from this call, ICAO established a Secretariat Study Group on Cybersecurity. It held its first meeting on 29 August 2017

154 States which are party to the PICASST: The *Cook Islands, Kiribati, Niue, Nauru, Papua New Guinea, Samoa, Solomon Islands, Tonga, Tuvalu, and Vanuatu.*
States which are not party to the PICASST: *Australia, Fiji and New Zealand.*
Associate member organisations: *Asian Development Bank, Association of South Pacific Airlines, Pacific Islands Forum Secretariat, and US FAA.*

155 ICAO, Recent Developments in Aviation Security, Muscat, Oman, 17 September 2017, DGCA-MID/4-IP/5.

and "developed a set of Terms of Reference and established several working groups with a view to addressing all elements and domains of the international aviation framework that may be affected by cyber incidents and attacks".[156]

The Secretariat Study Group on Cybersecurity falls under the responsibility of the Air Transport Bureau, so the focus of its work shall be on unlawful interference, rather than wider safety issues. In order fill this bureaucratic gap, under INNOVA – which is a network for the aviation stakeholders which provides a platform for global information exchange information in a trusted, secure and resilient environment – ICAO is in the process of establishing a Trust Framework Study Group. This will fall under the Air Navigation Bureau, and is likely to be enforced and supported during the 40th ICAO General Assembly.

6 Industry High Level Group

6.1 *Introductory Remarks*

In 2013, several key stakeholders – ACI, CANSO, IATA, ICAO and ICCAIA – established an IHLG. This was to create a mechanism for high-level cooperation on shared important issues in aviation security. This included cybersecurity issues and "the IHLG determined that cybersecurity in civil aviation was a high priority horizontal issue requiring aligned and coordinated actions by all relevant holders".[157] The work of the IHLG, which concentrates on the delivery and the promotion of the industry position on cybersecurity, then set the ground work which was taken up by the 39th ICAO Assembly.[158]

On 5 December 2014, the IHLG signed the Civil Aviation Cybersecurity Action Plan which outlines four declarations regarding aviation cybersecurity.[159] The undersigned organisations declare that they:

156 ICAO, Recent Developments in Aviation Security, Muscat, Oman, 17 September 2017, DGCA-MID/4-IP/5.
157 ICAO, Assembly – 39th Session, Executive Committee, Addressing Cybersecurity in Civil Aviation, Presented by the Council of ICAO, WA39-WP, EX/5, 30 May 2017.
158 IATA, 'Fact Sheet Cyber Security', https://www.iata.org/pressroom/facts_figures/fact_sheets/Documents/fact-sheet-cyber-security.pdf.
159 ICAO, Civil Aviation Cybersecurity Action Plan, 5 December 2014, Available: https://www.icao.int/cybersecurity/SiteAssets/ICAO/Civil%20Aviation%20Cybersecurity%20Action%20Plan%20-%20SIGNED.pdf.

- recognize the need to work together, guided by a shared vision, strategy and roadmap to strengthen the aviation system's protection and resilience against cyber-attacks;
- will jointly make the necessary effort to prepare civil aviation against future challenges from cyber threats;
- will cooperate on the following commitments:
  - Develop a common understanding of cyber threats and risks;
  - Share assessments of risks;
  - Agree common language and terminology;
  - Develop joint positions and recommendations;
  - Present a coherent approach to the public;
  - Promote cooperation among State-level appropriate authorise and industry to establish coordinated aviation cybersecurity strategies, policies and plans;
  - Promote a robust cybersecurity culture in all organizations in civil aviation;
  - Promote the use of existing information security and cyber protection best practices, standards and design principles, and establish new ones, where necessary;
  - Establish the mechanisms and means to share and communicate information including identification of threats, reporting of incidents and developments in defences;
  - Communicate threat-related information and assure situational awareness;
  - Refine best practices, operational principles and defensive systems, as appropriate.
- Support the actions defined in the roadmap attached to this document.

The scope of coordination and cooperation is vast, but it must be noted that this Action Plan does not create any legally binding obligations. Further, there is no obligation to share any confidential or sensitive information, which within the scope of cybersecurity, will be relevant.

One deliverable that the IHLG assisted with was the Resolution A39-19.

The rest of this subchapter will be dedicated to looking at some of the steps taken by the members of the IHLG.

6.2 *Airports Council International – World*

ACI was established in 1991 and represents airports' interests with governments and international organisations, develops standards, policies and recommended practices for airports, and it provides information and training opportunities to raise the standards around the world. It aims to provide the public with a safe, secure, efficient and an environmentally responsible air transport system.

ACI has developed an IT Security Benchmarking Tool which will assist airports develop and maintain a cybersecurity management programme which meets established international guidelines. The Tool was created under the management of the ACI World IT Standing Committee and the Cybersecurity Task Force.[160] As the Tool focuses on cybersecurity governance, management and oversight, its use is intended for use by the airport management team and the IT director. This is an important step as cybersecurity discussions are often about airlines, whereas airports, which are often victims of cyberattacks, are overlooked.

6.3 *Civil Air Navigation Services Organization*

CANSO, representing Air Navigation Service Providers (ANSPs), has members responsible for supporting over 85 percent of world air traffic, and it is involved in information sharing and policy shaping.

CANSO's information sharing role has included cybersecurity and, in June 2014, it published its Cyber Security and Risk Assessment Guide.

> The purpose of this document is to provide air navigation service providers with an introduction to cyber security in air traffic management, including the cyber threats and risks and motives of threat actors, as well as some considerations to managing cyber risks and implementing a cyber security programme.[161]

This is based on information sharing to make ANSPs aware of the cyber threats, cybersecurity measures, international standards and risk assessments. The Cyber Security and

160 Airport Cooperative Research Program, Guidebook on Best Practices for Airport Cybersecurity, ACRP REPORT 140.

161 CANSO, 'Cyber Security and Risk Assessment Guide', June 2014, https://www.canso.org/sites/default/files/CANSO International Organisations:CANSO%20Cyber%20Security%20and%20Risk%20Assessment%20Guide.pdf.

Risk Assessment Guide constitutes a practical guide for improving cybersecurity awareness across the ATM industry.[162] CANSO raised the profile of the Guide during the 39th General Assembly in a Working Paper that it is presented.[163]

6.4 *Eurocontrol*

6.4.1 Background

Eurocontrol is an international organisation working to on safe and seamless air traffic management across its European Member States.

6.4.2 System Wide Information Management

The System Wide Information Management (SWIM) concept "consists of standards, infrastructure and governance to facilitate the management of ATM-related information and its exchange between qualified parties through interoperable services".[164] SWIM "supports ATM operational activities by enabling the secured exchange of aeronautical, flight, meteorological, air traffic flow and surveillance information".[165] Eurocontrol has included cybersecurity within the scope of SWIM and has produced a brochure focusing on cybersecurity for SWIM implementation. The brochure puts forward questions that ATM providers should ask in the areas of service risk management; service development and evolution; service access management; service level agreements; information classification; ICT system protection; data protection; authentication, authorisation and non-repudiation; and operational management. These are based on different standards.[166]

6.4.3 CS6-7 Operations and Coordination of Network Security

Eurocontrol's "CS6-7 (Operations and Coordination of Network Security) is a service aimed towards coordinating cyber-security issues, consolidating ATM network cyber-security events/incidents and recommending actions when and if needed. This will strengthen the cyber-security of European ATM systems, reducing the risk of ATM/

162 Eurocontrol, 'Position Paper on Cybersecurity', http://www.iata.org/policy/Documents/cyber-threat-position.pdf.

163 ICAO, Working Paper, Assembly – 39th Session, Executive Committee, 'Cyber Security Defense Strategy', A39-WP/1871, EX/72, 10 August 2016.

164 ICAO, Manual on System Wide Information Management (SWIM) CONCEPT, Doc. 10039, AN/511.

165 Eurocontrol, 'SWIM Cybersecurity Hints & Tips', http://www.eurocontrol.int/publications/swim-cybersecurity-hints-tips.

166 *See*, ISO/IEC 27001:2013, Information technology – Security techniques – Information Security Management Systems – Requirements, https://www.iso.org/standard/54534.html; ISO/IEC 27002:2013, Information Technology – Security Techniques – Code of Practice for Information Security Controls, https://www.iso.org/standard/54533.html; and National Institute of Standards and Technology Special Publication 800-53 Revision 4 Security and Privacy Controls for Federal Information Systems and Organizations, http://www.nvlpubs.nist.gov/nistpubs/SpecialPublications/NIST.SP.800-53r4.pdf.

ANS service disruption or system corruption/compromise".[167] The CS6-7 has two main service categories:

1. EATM-CERT (EATM Computer Emergency Response Team).
2. Security Operations Centre (SOC).

First, on 13 March 2017 Eurocontrol and the CERT-EU signed a Service Level Agreement to improve cybersecurity. Under this Agreement, "CERT-EU will provide services to Eurocontrol in order to support the setting up of the European Air Traffic Management Computer Emergency Response Team (EATM-CERT). CERT-EU will make available some information and cyber-security tools to EATM-CERT as well as sharing experience, expertise, advice and procedures on how to operate a CERT efficiently".[168] EATM-CERT will provide the following services to ANSPs and Airport Operators of Eurocontrol Member States:

> Proactive cyber-security services (e.g. penetration testing, vulnerability assessment, infrastructure/best practices review);
> Collection, generation and distribution of ATM relevant cyber intelligence;
> Coordination, with National CERTs as appropriate, of the pan-European ATM response to ATM relevant cyber-security alerts/incidents on a voluntary basis.[169]

This is then a three-pronged approach to cybersecurity as it will strengthen defences with proactive measures, it will increase knowledge distribution so different stakeholders can benefit and it will be reactive as it will coordinate with national CERTs. This is a practical approach to cybersecurity outside the specific aviation regulations, however, the context of the EU NIS Directive cannot be ignored nor the CERT efforts by EASA.[170]

Second, a Security Operations Centre (SOC) can be defined as "a centralised unit that deals with security issues on an organisational and technical level".[171] For Eurocontrol, a SOC can fulfil several activities.

---

167 Eurocontrol, 'Centralised Service 6-7 Operations and Coordination of Network Security, Ensuring the Resilience of Centralised Services' Cyber-Security and Sharing Cyber Intelligence', February 2017.

168 Eurocontrol, 'Agreement with the European Commission's CERT-EU on Cyber-Security', http://www.eurocontrol.int/press-releases/agreement-european-commission-cert-eu-cyber-security.

169 Eurocontrol, 'European Air Traffic Management Computer Emergency Response Team (EATM-CERT)', http://www.eurocontrol.int/eatm-cert.

170 Eurocontrol, 'Centralised Service 6-7: EATM-CERT and SOC, Centralised Service 6-6: PKI, Presentation World Aviation Congress (WAC)', 8 March 2017, https://www.eurocontrol.int/sites/default/files/events/presentation/cs-6-6-and-6-7-briefing-presentation-day-2-eurocontrol-wac-2017.pdf.

171 A. Haghighi, 'An Introduction to SOC (Security Operation Center)', Presentation Published on 3 June 2016.

1. Analyse abnormal situations detected by SIEM (Security Information and Event Management) tool.
2. Filter false positive alerts (e.g. within 45').
3. Analyse true alerts/incidents and propose remediation actions (e.g. within 1 or 24 hours).
4. Improvement of abnormal situation detection criteria/threshold ('correlation rules') using CERT info.
5. Update SIEM with info provided by CERT.[172]

This can be either conducted by a central centre based on specific bilateral agreements or by the ANSP individually.

6.4.4 White Hack Hackathon

Between 4 and 6 November 2016, Eurocontrol held a 'hackathon' which brought 32 students to the Eurocontrol's training Institute of Air Navigation Services in Luxembourg. "The goal was to make use of their out-of-the-box thinking to help create the world's most innovative and efficient codec for ASTERIX (the All-purpose structured Eurocontrol surveillance information exchange)."[173] The contest was won by BlueCloud from Italy and each team member was awarded 1,000 EUR and an internship at Eurocontrol's headquarters. This novel approach by Eurocontrol not only provided it with a solution for a codec for ASTERIX but also provided it with a continuing benefit as it formed a competence-based recruitment. Aviation stakeholders will need to invest in staff or outsource professionals with IT skills, in order to introduce effective cybersecurity measures and this is a useful way of achieving this.

6.4.5 Training

Eurocontrol is also involved in cybersecurity training. For example, its course on Cyber Security in ATM [SEC-CYBER] "describes the Cyber threats to ATM, the vulnerabilities to be mitigated and the methods for managing incidents".[174]

172 P. Mana, Eurocontrol, 'Cyber-security Project Manager GANIS/SANIS', Presented at ICAO, Montreal, Canada, 11-15 December 2017.

173 Eurocontrol, 'The Eurocontrol Hackathon – Bringing Disruptive Technologies Through Out-of-the-Box Thinking', http://www.eurocontrol.int/news/eurocontrol-hackathon-bringing-disruptive-technologies-through-out-of-the-box-thinking.

174 Eurocontrol, 'Cyber Security in ATM [SEC-CYBER]', https://trainingzone.eurocontrol.int/ilp/pages/coursedescription.jsf?catalogId=4728296&courseId=6450279.

6.5 International Air Transport Association

6.5.1 Background

IATA is a trade association that represents more than 260 airlines from 120 countries' airlines amounting to 84 percent of total air traffic.[175] IATA's mission is to represent, lead and serve the airline industry.[176] First, it represents the airline industry by improving "understanding of the air transport industry among decision makers and increase awareness of the benefits that aviation brings to national and global economies. Advocating for the interests of airlines across the globe, we challenge unreasonable rules and charges, hold regulators and governments to account, and strive for sensible regulation".[177] Second, it leads the airline industry, as through the past "70 years, we have developed global commercial standards upon which the air transport industry is built [with the aim of assisting] airlines by simplifying processes and increasing passenger convenience while reducing costs and improving efficiency".[178] Finally, it serves the airline industry in helping it "to operate safely, securely, efficiently, and economically under clearly defined rules. Professional support is provided to all industry stakeholders with a wide range of products and expert services".[179]

6.5.2 Three-Pillar Strategy

In 2013, IATA "developed a three-pillar strategy to understand, define and assess the threats and risk of cyberattacks, the basis for appropriate regulation and the mechanisms for increased cooperation throughout the industry with the support of governments".[180] The three pillars are:

1. Risk Management:
   a. Develop a framework for and provide input to Threat Guide for cybersecurity in aviation, and
   b. Develop a Cyber Security Management System to assist airlines in performing risk assessment and implementing mitigation strategies.
2. Advocacy:
   a. Advocate with key regulators for risk-based regulation and coordination requirements, and
   b. Support airline leadership and technicians in advocating 'security by design' measures with systems providers and manufactures.
3. Reporting and Communication:

175 IATA, 'About Us', http://www.iata.org/about/pages/index.aspx.
176 IATA, 'Vision and Mission', http://www.iata.org/about/Pages/mission.aspx.
177 *Ibid.*
178 *Ibid.*
179 *Ibid.*
180 IATA, 'Fact Sheet Cyber Security', https://www.iata.org/pressroom/facts_figures/fact_sheets/Documents/fact-sheet-cyber-security.pdf.

a. Raise awareness of cybersecurity issues through outreach and communication programmes, and
b. Identify reporting and intelligence sharing mechanisms.

IATA proposed the following solutions:

> Continually update its position on cybersecurity to represent the interests of its members.
>
> Develop new venues for cooperation with stakeholders and entities to work on cyber mitigation measures for the aviation industry.
>
> Assist its members in maintaining and sharing best practices in relation to cyber mitigation and resilience and business continuity.
>
> Support continuous monitoring and assessment of the potential risks, particularly in relation to the prevention of unlawful interference against aircraft.
>
> Proactively advocate to governments and industry regulators smarter regulations and the prevention of excessive or prescriptive legal frameworks that would challenge industry capabilities to overcome the cyber threat.[181]

Such assistance includes providing cybersecurity knowledge to mitigate threats through its Security Management System (SeMS).[182]

6.5.3 Regulatory Efforts

IATA takes two core roles in regulatory efforts for safety and security. Firstly, it "encourages governments to use a harmonized approach to regulation".[183] Secondly, it works with ICAO within its role of developing SARPs. In its work, IATA has shelved cybersecurity under the heading of security.

6.5.4 Cybersecurity Toolkit

IATA has assisted its member airlines in developing, implementing and enhancing their cybersecurity programmes through the Cybersecurity Toolkit.[184] In 2015, IATA published the second edition of the Aviation Cyber Security Toolkit which assists airlines

181 IATA, 'Position Paper on Cybersecurity', http://www.iata.org/policy/Documents/cyber-threat-position.pdf.
182 IATA, 'Insider Threat in Civil Aviation, Position Paper for 2015/2016', https://www.iata.org/policy/Documents/insider-threats-position.pdf.
183 IATA, 'Safety & Security', http://www.iata.org/policy/safety-security/Pages/index.aspx.
184 IATA, 'Position Paper on Cybersecurity', http://www.iata.org/policy/Documents/cyber-threat-position.pdf.

in raising awareness and understanding, as well as better defining the cyber risks to which airlines are exposed to. The toolkit includes:

1. Part 1 – Covers an overview of the current situation.
2. Part 2 – Introduces cybersecurity threats.
3. Part 3 – Outlines cybersecurity risk mitigation practices.
4. Part 4 – Provides guidance for the implementation of a cybersecurity management system.
5. Part 5 – Gives access to a workbook for organisations to make their own robust risk assessments as well as prioritisation instructions.

IATA also cooperates with other stakeholders. For example, it has been involved with the IHLG and its Roadmap, and the proposed declaration on cybersecurity that was developed and adopted by the 39th ICAO Assembly.[185]

In 2017, it published its partnership with the Aviation Information Sharing and Analysis Center (A-ISAC) "to promote a tiered membership programme that enables all aviation service providers to engage in A-ISAC's premier global cybersecurity threat sharing network, regardless of size or revenue".[186] This cooperative action will widen the stakeholder's ability to engage in threat sharing and analysis capabilities.

6.5.5 Operations Work Groups

IATA's Operations Work Groups develop operational solutions for all areas affecting aircraft and airlines, either on ground or in flight. The Security Group, which is one of the Operations Committee, meets twice a year and incorporates five underlying pillars in its work:

1. Collaboration with government and industry.
2. Threat based, outcome focused approach.
3. Regulatory Framework.
4. Cost and efficiency.
5. Innovation and technology.[187]

Key to the Security Group's contributions are recommendations and projects that improve the value proposition for aviation security and facilitation and focus on threat and

185 ICAO, Assembly – 39th Session, Executive Committee, Addressing Cybersecurity in Civil Aviation, WA39-WP, EX/5, 30 May 2017.

186 Airport Improvement, 'Aviation ISAC Announces New Membership Structure and Partnership IATA Strengthens Rapid International Growth', http://www.airportimprovement.com/news/aviation-isac-announces-new-membership-structure-and-partnership-iata-strengthens-rapid-international-growth.

187 IATA, 'Security Group (SEG)', http://www.iata.org/whatwedo/workgroups/Pages/secwg.aspx.

risk assessments and mitigation in areas, such as:

1. Passenger and cargo risk-based screening.
2. Data driven security.
3. SeMS and aircraft operator security programs.
4. Cybersecurity.
5. Emerging threats.
6. Crisis Management and operational contingency planning.[188]

As a result, IATA is focused on cybersecurity and has implemented several activities to help its member airlines with this issue.

6.6 *International Coordinating Council of Aerospace Industries Associations*

The ICCAIA was established in 1972 in order to provide aerospace manufacturers representation at ICAO via observer status. It is currently made up of six members.

1. AeroSpace and Defence Industries Association of Europe.
2. Aerospace Industries Association of America (US).
3. Aerospace Industries Association of Brazil.
4. Aerospace Industries Association of Canada.
5. Society of Japanese Aerospace Companies.
6. Union of Aviation Industrialists (Russia).

The majority of ICCAIA's work is carried out in its four committees composed of staff from its six member associations and the companies belonging to the member associations. The Security Committee is responsible for cybersecurity within the ICCAIA. Apart from being a member of the IHLG and producing General Assembly Working Papers, its members are active within ICAO Panels and Study Groups.

6.7 *International Federation of Air Line Pilots*

6.7.1 Background

IFALPA, while not a member of the IHLG, is also worth exploring. It is a worldwide association of national pilots' associations. IFALPA's mission "is to promote the highest level of aviation safety worldwide and to be the global advocate of the piloting profession; providing representation, services and support to both our members and the aviation

188 *Ibid.*

industry".[189] Its mission is to be achieved via core functions which are "[i]nteracting with international organizations to achieve the highest level of aviation safety; [p]romoting and enhancing the role and status of professional pilots in ensuring the safety of the aircraft and well-being of passengers and goods entrusted to their care; and [p]roviding Member Associations with services as needed".

6.7.2 Position Paper

IFALPA, has highlighted its concerns regarding cyber threats to aircraft, ground facility, and other critical infrastructure in a Position Paper.[190] In the document, "some guidelines are provided that could help in establishing an environment in which this threat is fully understood and managed and the risk has been brought down to an acceptable level".[191] These guidelines first note that "States should establish regulations to set the minimum requirements that the aviation industry should meet".[192] These rules should consist of technical requirements, contingency planning, proper training of all personnel and compliancy auditing. Second, stakeholders should share information "on security breaches, detected attacks and best practices", within an environment of trust, as this will benefit cybersecurity.

6.7.3 Security Briefing Leaflet

IFALPA supplemented the Position Paper with a Security Briefing Leaflet.[193] This provides guidelines to assist stakeholders "in establishing an environment in which this threat is fully understood and managed and the risk has been brought down to an acceptable level".[194] This takes a six-tier approach focusing on software, hardware, data protection (electronic and physical), training, governance and control, and information sharing. Nothing here is ground-breaking, but it shows consistency between stakeholders on the approach to be taken with regard to aviation cybersecurity.

7 International Criminal Aviation Law Treaties

7.1 Introductory Remarks

A specific reference is made to the international criminal aviation law treaties – notably; the Tokyo Convention 1963, The Hague Convention 1970 and the Montreal Convention

189 IFALPA, 'Mission Statement', https://www.ifalpa.org/about-us/mission-statement.html.
190 IFALPA, 'Cyber Threats, Position Paper 16POS08', 6 December 2016.
191 IFALPA, 'Cyber Threats, Security Briefing Leaflet 16SECBL01', 6 December 2016.
192 IFALPA, 'Cyber Threats, Position Paper 16POS08', 2016.
193 IFALPA, 'Cyber Threats, Security Briefing Leaflet 16SECBL01', 6 December 2016.
194 *Ibid.*

1971 – in Regulation 300/2008 due to their relevance in dealing with prohibited activities against international civil aviation. These Conventions are also useful tools for the interpretation of Article 3 *bis*, the scope of security-based articles in ASA and for placing Annex 17 within its wider international law context. As a result, the international criminal aviation law treaties must be examined to fully address the regulation of aviation cybersecurity at an EU level.

The delegates at the Chicago Conference in 1944 did not include provisions to regulate prohibited activities of passengers and others on the ground, such as physical hijacking and sabotage. However, they were not unaware of the importance that the abuse of international civil aviation could have on the world. For example, the Preamble states:

> WHEREAS the future development of international civil aviation can greatly help to create and preserve friendship and understanding among the nations and peoples of the world, yet its abuse can become a threat to the general security.

Aviation security began to take on a practical meaning in the 1950s with the increasing number of "extremely violent terrorism against aviation and the unlawful seizure of aircraft".[195] This forced criminal air law to become a key area of ICAO's activities. This subchapter will examine the treaty developments done through the auspices of ICAO and the Budapest Convention to see how cybersecurity is promoted and cyberattacks are regulated.

7.2 *Cybersecurity Offences*

7.2.1 Tokyo Convention 1969

Between 25 August and 12 September 1953, the ICAO Legal Committee officially established a subcommittee on The Legal Status of Aircraft to study the issues connected with crimes on board aircraft. A draft convention was developed in 1958. The final draft was prepared in 1962, finalised and then adopted by the Diplomatic Conference which convened in Tokyo, Japan by the ICAO Council from 20 August to 14 September 1963. This resulted in the Convention on Offences and Certain Other Acts Committed on Board

195 ICAO, 'The Postal History of ICAO: Legal Instruments Related to Aviation Security', http://www.icao.int/secretariat/PostalHistory/legal_instruments_related_to_aviation_security.htm.

Aircraft, which is more commonly known as the Tokyo Convention.[196] It entered into force on 4 December 1969 and currently has 186 Parties (as of July 2019).[197]

Prior to the Tokyo Convention, many offences committed on board aircraft would go unpunished due to a lack of jurisdiction during international flights.[198] The Convention was drafted in order to find a solution to the 'oases of lawlessness' that aircraft used to be in when they were flying in the territory of other States or especially above the high seas.[199] Therefore, the primary concern of the Convention is to establish jurisdiction to offences and acts committed on board any aircraft registered in a contracting State, while that aircraft is in flight or on the surface of the high seas or any other area outside the territory of any State.[200]

Article 1 clarifies which offences are covered by the Convention.

> This Convention shall apply in respect of:
> a. offences against penal law;
> b. acts which, whether or not they are offences, may or do jeopardize the safety of the aircraft or of persons or property therein or which jeopardize good order and discipline on board.

Article 1(1)(a) does not specify conduct that is prohibited by the Tokyo Convention. Rather, it is left to domestic penal law to determine what constitutes an offence. Many States involved in international aviation have penal law prohibiting certain acts, which could include conduct on board an aircraft, so this provides contracting States with high levels of autonomy to regulate this area. As such, penal law covering cyberattacks may extend, or could be introduced or amended to cover acts on board of aircraft.

Article 1(1)(b) makes it an offence for a person to jeopardise safety, good order and discipline on board an aircraft even when this is not prescribed in that State's domestic penal law. This has an objective element where the acts 'do', but also a subjective element where the acts 'may', jeopardise the safety or good order and discipline on board of an

196 Convention on Offences and Certain Other Acts Committed on Board Aircraft, UNTS, 1969, No. 10106, Vol. 704, 14 September 1963, at 216. (Tokyo Convention.)

197 ICAO, 'Convention on Offences and Certain Other Acts Committed on Board Aircraft Signed at Tokyo on 14 Sep. 1963', http://www.icao.int/secretariat/legal/List%20of%20Parties/Tokyo_EN.pdf.

198 *See*, *US v. Cordova*, 89 F. Supp. 298 E.D.N.Y. (1950); and *R. v. Martin* 2 Q.B. 272 (1956).

199 B. Cheng, 'The Convention of Tokyo, The Hague, Montreal and a new instrument concerning unlawful violence at international airports', *in* Tanya Masson-Zwaan, *Aviation Security Law Reader, Leiden University* (2013), at 19-30.

200 P. van Krieken, 'Hijacking and Asylum: Some Remarks on the Hijacking Problem and its Repression, with Respect to the Law on Asylum', 22(1) *Netherlands International Law Review* 3-30 (1975), at 8.

aircraft. The latter then only requires the act to be a perceived jeopardy. As the Tokyo Convention does not provide an exhaustive list of definitions, such as with regard to the terms 'jeopardise', 'safety', 'good order' and 'discipline', the test for assessing Article 1(1)(b) is unclear with "its definition is to be determined by the municipal laws of the contracting States".[201] Therefore, if an attacker commits a cyberattack on board an aircraft, in a way that satisfies section (b), then the Convention will be applicable.

The Convention is limited in relation to the scope of cyberattacks.

1. It only applies to acts or offences that occur 'on board of the aircraft', so they do not include cyberattacks against aviation stakeholders from all locations.
2. It only applies to acts or offences which occur while the aircraft is 'in flight', so the whole operation is not covered, such as acts committed in the airport or while the doors of the aircraft are open.[202]
3. Pursuant to Article 2, offences of a "political nature or those based on racial or religious discrimination" do not authorise or require any action from States that have jurisdiction. Past examples of cyberattacks in aviation have been politically motivated, so this could prevent some cyberattacks going unprosecuted.
4. Finally, nothing in the Convention creates an obligation to extradite nor prosecute the offender. Therefore, this could prevent some cyberattacks going unprosecuted. As it only applies to such acts or offences on board an aircraft, the extradition element issue is lessened for cyberattacks with can be implemented anywhere in the world.

This can leave passengers and airlines with weak protection at an international level and will result in matters outside the scope of the Convention being dealt with under domestic law.

7.2.2 The Hague Convention 1970

The Tokyo Convention marked the first international effort to criminalise certain acts against international civil aviation. However, it failed to fully address the issue of hijacking, which saw a big increase in the 1960s and 1970s. Early examples were focused around the Middle East and the Caribbean, but as the 9/11 attacks show, no part of the world is immune.[203] As the number of hijackings rose from 1968 to 1970, further legal measures were required, thus from 1 to 16 December 1970, 77 States and 12 international organi-

---

201 A. Evans, Aircraft Hijacking: Its Cause and Cure, 63(4) *American Journal of International Law* 695-710 (1969), at 708.

202 Tokyo Convention, Art. 1(3):

> an aircraft is considered to be in flight from the moment when power is applied for the purpose of take-off until the moment when the landing run ends.

203 *See*, as a detailed description of The Hague Convention, Abraham Abramovsky, 'Multilateral Conventions for the Suppression of Unlawful Seizure and Interference with Aircraft Part I: The Hague Convention', 13 *Columbia Journal of Transnational Law* 381-405 (1974).

sations met in The Hague, The Netherlands for a diplomatic conference. The result was the Convention for the Suppression of Unlawful Seizure of Aircraft (The Hague Convention or Hijacking Convention),[204] which entered into force on 14 October 1971 and currently has 185 State Parties (as of July 2019).

The Convention acts to supplement the Tokyo Convention in such a way so as to establish criminal jurisdiction of the State of registry over more serious offences committed on board aircraft; e.g. hijacking. Article 1(a) states:

> Any person who on board an aircraft in flight:
> a. unlawfully, by force or threat thereof, or by any other form of intimidation, seizes, or exercises control of, that aircraft, or attempts to perform any such act,
> b. is an accomplice of a person who performs or attempts to perform any such act commits an offence (hereinafter referred to as 'the offence').

Four constituent elements can be found in Article 1(a).

1. There must be a 'use of force', threat or other form of intimidation in the act.
2. The act must involve 'seizure, exercise of control' of that aircraft or an attempt to do such an act.
3. The aircraft must be in flight.
4. It is only applicable to acts committed 'on board of the aircraft', so it will not include a remote cyberattack, such as spoofing which can amount to hijacking.

If an act satisfies the above criteria, then it is unlawful and referred to as 'the offence' for the purpose of the Convention. If the Roberts case is substantiated,[205] then this would show that there is a possibility of gaining unauthorised access and controlling an aircraft while on board. This clearly satisfies points 2 and 3. However, it is not so clear if point 1 is satisfied. While the use of force could include "the use of physical force, firearms or other weapons, as well as the threat to use such measures of coercion",[206] it is not clear if this would include sitting quietly, using a computer with no one being aware, including the pilots on board of the attacked aircraft. The *travaux préparatoires* of the Legal Committee of ICAO appears to answer this in the negative as it seems to only include actual use of force. For example, the

204 Convention for the Suppression of Unlawful Seizure of Aircraft (The Hague Convention), The Hague, UNTS, 1973, No. 12325, Vol. 860, at 106, 16 December 1970.

205 *See*, Chapter I.1.

206 S. Shubber, 'Aircraft Hijacking under the Hague Convention 1970 – A New Regime?', 22 *International and Comparative Law Quarterly* 687-726 (1973), at 690.

> Australian delegate proposed the enlargement of the scope of the provision in question to include the exercise of unlawful control over aircraft 'in any other manner' so that seizure of aircraft that is carried out through trickery, false pretence or bribery of the pilot could be covered.[207]

The proposal was rejected by the Legal Committee by 25 votes to seven.[208] This is encapsulated by France's statement that "[o]ne of the conditions of the offence should be the use of force, whether this was of a moral or physical nature. If there was no coercion, the offence would extend to cases where there was an agreement between the pilot and the offender 'for unlawful seizure of the aircraft and this would not jeopardize the' safety of the aircraft".[209] Spain made a similar statement.[210] Therefore, without such use of force, which was absent in the Roberts case, it is unlikely that a cyberattack would invoke The Hague Convention. The Convention may, however, be applicable for cyberattacks that use force and it may be the case that judicial interpretations place such acts under the scope of the Convention. If this is the case, then pursuant to Article 2 each contracting State shall "make the offence punishable by severe penalties".

7.2.3 Montreal Convention 1971

The Convention for the Suppression of Unlawful Acts Against the Safety of Civil Aviation (Montreal Convention 1971 or the Sabotage Convention)[211] was a reaction to a number of cases. The first case took place on 21 February 1970 and was in Switzerland. Bombs were placed in postal packages on Swissair Coronado Flight 330 flying from Zurich, Switzerland to Hong Kong via Tel Aviv, Israel. The bombs exploded and tragically killed all 47 people on board. The second case took place on the same day and involved a bomb being placed on board an Austrian Airlines aircraft flying from Frankfurt, Germany to Vienna, Austria. The pilots were able to successfully land the aircraft without any fatalities. The perpetrators of the two events were never caught,[212] but the Popular Front for the Liberation of Palestine claimed responsibility for both cases. It became clear to the international community and aviation stakeholders that there was a need for an international treaty dealing with sabotage. The Montreal Convention 1971 entered into force on 26 January 1973 and currently has 188 State Parties (as of July 2019).

---

207 ICAO, Report of the Legal Committee, 17th Session, on the Subject of Unlawful Seizure of Aircraft, Minutes and Documents, ICAO Doc. 8877-LC/161, at 28.
208 *Ibid.*, at 33.
209 *Ibid.*, at 29.
210 *Ibid.*, at 29.
211 Montreal Convention 1971.
212 Aviation Safety Network, 'Saturday 21 February 1970', https://aviation-safety.net/database/record.php?id=19700221-1.

The nature of the offences targeted by the Convention are attacks on board of an aircraft, and acts of sabotage against an aircraft and air navigation facilities. Hence, it criminalises acts 'against' the aircraft, as opposed to the Tokyo and The Hague Conventions which apply to acts 'on board of the aircraft'. The material scope of application of the convention is set out in Article 1 of the Montreal Convention 1971, which reads as follows:

1. Any person commits an offence if he unlawfully and intentionally:
   a. performs an act of violence against a person on board an aircraft in flight if that act is likely to endanger the safety of that aircraft; or
   b. destroys an aircraft in service or causes damage to such an aircraft which renders it incapable of flight or which is likely to endanger its safety in flight; or
   c. places or causes to be placed on an aircraft in service, by any means whatsoever, a device or substance which is likely to destroy that aircraft, or to cause damage to it which renders it incapable of flight, or to cause damage to it which is likely to endanger its safety in flight; or
   d. destroys or damages air navigation facilities or interferes with their operation, if any such act is likely to endanger the safety of aircraft in flight; or
   e. communicates information which he knows to be false, thereby endangering the safety of an aircraft in flight.

Therefore, this Convention was mainly designed to cover many kinds of attacks. The act must be done both *unlawfully* and *intentionally* for it to be an offence.

1. Unlawfully: The act must be forbidden in law. As a result, this would exclude actions that are "legally justifiable or done with legal authority, such as preventive action by police".[213] It may also include measures adopted by airlines in the contract of carriage.
2. Intentionally: The act must be performed with intent; it must possess the *mens rea* element. As a result, accidental damage, is not included.

It is unlikely that Article 1(1)(a) is relevant in this case as cyberattacks do not normally amount to a violent act as this usually requires a physical act. However, where a cyberattack produces physical effects, then this could then become relevant.

213 Commonwealth Secretariat, 'Implementation Kits for the International Counter-Terrorism Conventions', https://www.unodc.org/pdf/crime/terrorism/, Ch. 4, at 77.

The violence must also endanger the safety of the aircraft. This can lead to certain acts falling outside the scope of the Montreal Convention 1971, such as intimidation, sexual assault[214] of passengers and crew,[215] or drunken passengers[216] as these may or may not affect the safety of the aircraft.[217] This opens the debate as, for example, it can be posited that "such acts do not directly endanger the safety of the aircraft and, hence, do not come within the ambit of the Convention".[218] On the other hand, the flight and cabin crew may be required to deviate from their normal duties to assist, which may impact upon safety.

The second offence is found in Article 1(1)(b) and includes destroying an aircraft in service[219] or causing to damage it to an extent that the flight would be impossible or unsafe. An example of this is the Malaysia Airlines MH17 which was shot down by a missile on 17 July 2014 while flying over Ukraine, tragically killing all 283 passengers and 15 crew members on board.[220] The term 'destroy' can be interpreted to mean that the act puts the aircraft in a position beyond repair.[221] The term 'damage' would result in an act that alters the aircraft from its prior State, but that is reparable. The line between the two may be grey in certain circumstances, however, as the Sabotage Convention does not prescribe different outcomes for the two, this distinction is not relevant. The test for 'endangering the safety of the flight' is not set in the Sabotage Convention. This could include real threats or even theoretical ones, such as cases that impact upon the aircraft's airworthiness in non-flight critical ways. This is an offence that could be committed against an aircraft by a cyberattack.

---

214 *Tsevas v. Delta Airlines*, No. 97 C 0320. N.D. Ill. (1997).

215 *Evgeny Dogaev (a.k.a. Evgeny Rotshtein) v. Czech Republic*, [2014] EWHC 2995.

216 *R v. Heather Susan Tagg*, [2001] EWCA Crim 1230.

217 *See*, V. Correia, 'Some Cases on Unruly Behaviour and Criminal Offences Committed on Board Aircrafts', University of Poitiers, France, https://www.academia.edu/23558275/Some_cases_on_unruly_behaviour_and_criminal_offences_committed_on_board_aircrafts_-_Beijing_9_April_2015.

218 A. Abramovsky, 'Multilateral Conventions for the Suppression of Unlawful Seizure and Interference with Aircraft Part II: The Montreal Convention', 14 *Columbia Journal of Transnational Law 268-300* (1975), at 285.

219 Montreal Convention, Art. 2(b):

> An aircraft is considered to be in service from the beginning of the preflight preparation of the aircraft by ground personnel or by the crew for a specific flight until twenty-four hours after any landing; the period of service shall, in any event, extend for the entire period during which the aircraft is in flight as defined in para. (a) of this Article.

220 *See*, P. Olcer, 'The Downing of MH17: Fact-Finding and Constructing a Legal Response', 42(2) *Air and Space Law* 137-162 (2017); and Paul Freeman, Melissa Tang and Lau Chee Chong, 'Lightning Doesn't Strike Twice: The Tragedy and Challenges of the MH370 and MH17 Losses', 42(4/5) *Air and Space Law* 423-442 (2017).

221 S. Jung, 'A Legal Analysis of Aviation Security under the International Legal Regime', Institute of Air and Space Law, *McGill University* (2005), at 32.

Pursuant to Article 1(1)(c), the placing of a device or substance on an aircraft in service[222] that is likely to destroy or cause damage to it in a way that would render it incapable of flying or endanger flight safety is considered to be an offence under the Convention.

1. The act can be 'direct' or 'indirect'.
   1.1 Direct: This can mean physically taking the device or substance on board as a passenger or crew, or delivering it in person by placing in on board.
   1.2 Indirect: This is where the individual uses a third person to assist. This can include sending the device or substance by mail or using a cargo company to transport it.
2. The phrase 'by any means whatsoever' can include the placing on board and on the outside of the aircraft.
3. The item in question must either be a 'device' or 'substance'.
   3.1 Device: This refers to something that is made, such as mechanical or electronic equipment.
   3.2 Substance: This refers to matter, such as chemicals.
   3.3 Weapon: This term was not included, as is the case in the additional treaties, and the inclusion of the term was in fact rejected.[223]

If the above is satisfied, then the act is an offence if it destroys the aircraft, damages the aircraft making it incapable of flight, or endangers safety in flight. This could apply to aircraft which have been subject to a cyberattack. However, it must relate to the destruction of the aircraft or damage which renders it incapable of flying. The Convention does not specifically define what has to be understood as a 'substance' or 'device'; hence this could be interpreted very broadly.

Article 1(1)(d) is the first move away from the aircraft itself as it makes it an offence to destroy, damage or interfere with air navigation facilities. The scope of the term 'air navigation facilities' is not defined for in the Convention. However, Article 4(5) provides an interpretive tool.

> In the cases contemplated in subparagraph (d) of paragraph 1 of Article 1, this Convention shall apply only if the air navigation facilities are used in international air navigation.

Consequently, "the scope of air navigation facilities is restricted to airports, radio and meteorological services used in international flights as stipulated in Article 28 (Air navigation facilities and standard systems) of the Chicago Convention".[224] Air navigation

222 Montreal Convention 1971.

223 ICAO, International Conference on Air Law (Convention for the Suppression of Unlawful Acts against the Safety of Civil Aviation), Doc. 9081-LC/170-1 (1971).

224 S. Jung, 2005, at 33. *See*, for a definition, Beijing Convention, Art. 2(c).

facilities have been the focus of some cyberattacks and, with their importance for the safe navigation of civil aircraft through the airspace, it is clear that a cyberattack could endanger an aircraft in flight. Hence, it would indeed be possible for someone to destroy, damage or interfere with the air navigation facilities in a way that would endanger the safety of the aircraft.

Finally, Article 1(1)(e) makes it an offence to knowingly communicate false information that endangers the safety of an aircraft in flight. This can include bomb threats like the one in February 2016, where:

> Hello @Ryanair, you have 15 minutes before I commit the biggest terror attack the UK has ever seen on one of your planes. Be ready

was posted on Twitter.[225] The threat was vague enough to include all Ryanair's aircraft, so with regard to those in flight, the act could fall under the scope of the Sabotage Convention. There must be a link to safety and here two positions exist.

1. False bomb alerts, that cause only delay may remain outside the scope of the Sabotage Convention as there is objectively no direct link to safety.
2. False bomb alerts, which, for example, force the pilot to make an emergency landing could endanger the safety of the flight. Thus, there is a link between the bomb threat and safety.

The aircraft that were on the ground would not fall within the scope of the Convention as they were not in flight.

The Ryanair case involved the reporting of false information, which is covered by Article 1(1)(e). Spoofing is an attack method that involves reporting false information, so this offence could cover such acts. However, it may not cover jamming as this is the act of blocking the transfer of information. As a result, Article 1(1)(e) does not go far enough to cover all information exchanges.

Finally, in addition to not covering information blocking, it does not cover the transmission of real reports or mistaken reports.

225 S. Calder, 'Ryanair Wins £220,000 from Twitter User who Posted Terror Threat', The Independent, 14 July 2017, http://www.independent.co.uk/travel/news-and-advice/ryanair-bomb-threat-terror-twitter-brian-lake-pennsylvania-extortion-a7841446.html.

7.2.4 Montreal Protocol 1988

The Protocol for the Suppression of Unlawful Acts of Violence at Airports Serving International Civil Aviation (Montreal Protocol 1988)[226] supplements the Montreal Convention 1971 and, pursuant to Article I, both shall be read and interpreted together as one single instrument.[227] The Protocol entered into force on 6 August 1989 and has 175 State Parties (as of July 2019). As the Montreal Convention focuses on aircraft in flight and air navigation facilities, it was observed that "acts of violence which endanger or are likely to endanger the safety of persons at airports serving international civil aviation or which jeopardize the safe operation of such airports" are omitted from international conventions. Therefore, the Protocol was an international attempt to include airports in the legal regime.

Article II of the Protocol, which supplements Article 1 of the Montreal Convention 1971, adds the following offences as listed in the Montreal Convention:

> Any person commits an offence if he unlawfully and intentionally, using any device, substance or weapon:
> a. performs an act of violence against a person at an airport serving international civil aviation which causes or is likely to cause serious injury or death; or
> b. destroys or seriously damages the facilities of an airport serving international civil aviation or aircraft not in service located thereon or disrupts the services of the airport,
> if such an act endangers or is likely to endanger safety at that airport.

As highlighted above, a cyberattack will not necessarily amount to an act of violence so the relevance of (a) is currently limited within this context.

If the cyberattack destroys or damages facilities at an airport, then (b) will be applicable. This, again, produces a difficulty with definitions. For example, the Vietnam International Airport case of July 2016 resulted in information screens and sound systems being over-taken by unauthorised sources.[228] No destruction or physical damage occurred to any airport facilities as the attack happened remotely and, therefore, would not fall within the scope of the Protocol. Perhaps if some physical damage occurred as a result of a

226 Protocol for the Suppression of Unlawful Acts of Violence at Airports Serving International Civil Aviation, Supplementary to the Convention for the Suppression of Unlawful Acts against the Safety of Civil Aviation (Montreal Protocol 1988), UNTS 1990, 24 February 1988.

227 The Montreal Protocol 1988 is only binding on States that have ratified it, so it will not interact with States' obligations under the Montreal Convention 1971 if they are not State Parties to the Montreal Protocol 1988.

228 *See*, Chapter I.1.

cyberattack, such as with the Stuxnet attack, or if physical damage occurred in order to implement the attack, then the Protocol would be applicable.

7.2.5 Beijing Convention 2010

The Convention on the Suppression of Unlawful Acts Relating to International Civil Aviation (Beijing Convention)[229] was signed in Beijing on 10 September 2010. In part, it was a response to the 9/11 terrorist attacks against the US. However, not all of its scope covers the 9/11 facts, so as opposed to other Conventions, such as the Montreal Convention 1971, this Convention was not purely a reaction to a specific event, but was also a proactive response to some new threats facing international civil aviation. Although it mainly consolidates the Montreal Convention 1971 and the Montreal Protocol 1988, with some further amendments and additions to modernise the texts, its entry into force was not always certain and it currently has only 29 State Parties (as of July 2019), whereby Turkey's ratifications allowed the Convention to enter into force per Article 22.

Aside from the original offences set out in the Montreal Convention 1971 and the Montreal Protocol 1988, the Beijing Convention makes it an offence per Article 1 if someone unlawfully and intentionally:

(f) uses an aircraft in service for the purpose of causing death, serious bodily injury, or serious damage to property or the environment; or
(g) releases or discharges from an aircraft in service any BCN weapon or explosive, radioactive, or similar substances in a manner that causes or is likely to cause death, serious bodily injury or serious damage to property or the environment; or
(h) uses against or on board an aircraft in service any BCN weapon or explosive, radioactive, or similar substances in a manner that causes or is likely to cause death, serious bodily injury or serious damage to property or the environment; or
(i) transport, cause to be transported, or facilitate the transport of, on board an aircraft;
    1. any explosive or radioactive material, knowing that it is intended to be used to cause, or in a threat to cause, with or without a condition, as is provided for under international law, death or serious injury or damage for the purpose of intimidating a population, or compelling a government or an international organization to do or to abstain from doing any act;

229 Convention on the Suppression of Unlawful Acts Relating to International Civil Aviation (Beijing Convention), (Doc. 9960), 10 September 2010.

2. any BCN weapon (as defined in Article 2);
3. any source material, special fissionable material, or equipment or material especially designed or prepared for the processing, use or production of special fissionable material, knowing that it is intended to be used in a nuclear explosive activity or in any other nuclear activity not under safeguards pursuant to a safeguards agreement with the International Atomic Energy Agency;[230]
4. any equipment, materials or software or related technology that significantly contributes to the design, manufacture or delivery of a BCN weapon without lawful authorization and with the intention that it will be used for such purpose.

It may be the case that Article 1(f) is relevant for cyberattacks if an act results in an aircraft being used as a weapon, such as via spoofing, which would then be an offence under the Convention. However, no definition of 'weapon' is provided and the discussion in Chapter III.3.2 on the definition of 'weapon' can be used to highlight this point.

A cyberattack does not fall under the definition of biological, chemical and nuclear weapons (BCN), explosive, radioactive or similar substances. As a result, these fall outside the scope of the majority of new inclusions found in the Beijing Convention.

7.2.6 Beijing Protocol 2010

The Beijing Protocol[231] supplements The Hague Convention. The Protocol recently entered into force, pursuant to Article XXIII, with the Ratification of Uganda.[232] It currently has 32 State Parties (as of July 2019). The Beijing Protocol appears to address the use of force issue found in The Hague Convention. Article II declares:

> Any person commits an offence if that person unlawfully and intentionally seizes or exercises control of an aircraft in service by force or threat thereof,

230 This does not include activities involving or authorised by a State Party. For example, it will not be an offence if the transport of such items or materials is consistent with, or is for the use or for an activity that is consistent with the rights, responsibilities and obligations under the applicable multilateral international treaty law. Notable examples include the Treaty on the Non-proliferation of Nuclear Weapons, 729 UNTS 161; 7 ILM 8809 (1968); 21 UST 483, 1 July 1968; the Convention on the Prohibition of the Development, Production and Stockpiling of Bacteriological (Biological) and Toxin Weapons and on Their Destruction, 1015 UNTS 163; 11 ILM 309, 10 April 1972; and the Convention on the Prohibition of the Development, Production, Stockpiling and Use of Chemical Weapons and on Their Destruction, 1974 UNTS 45; 32 ILM 800, 13 January 1993.

231 Protocol Supplementary to the Convention for the Suppression of Unlawful Seizure of Aircraft (Beijing Protocol), ICAO Doc. 9959 (2010).

232 ICAO, 'Beijing Protocol to Enter into Force Following Uganda's Ratification', https://www.icao.int/Newsroom/Pages/Beijing-Protocol-to-enter-into-force-following-Ugandas-ratification.aspx.

or by coercion, or by any other form of intimidation, or by any technological means.

The expression 'any technological means' was added in an attempt "to address situations, in which the offenders seek to take control of an aircraft by jamming or otherwise interfering with flight instruments or data transmission systems".[233] The wording is extremely broad to include all cyberattacks. However, this must be read in the light of the rest of the Article and The Hague Convention, so it has the same general limitations as highlighted above. For example, the Boeing 757 hack example, whereby an aircraft was controlled from a remote location, would not be included as it was not 'in service'.[234]

7.2.7 Montreal Protocol 2014

Unruly and disruptive passenger events are a real and continuing concern for airlines worldwide and there has been a need to amend the Tokyo Convention. Therefore, from 26 March to 4 April 2014, under the auspices of ICAO, an International Conference on Air Law was held in Montreal to consider amending the Tokyo Convention. This resulted in the Protocol to amend the Convention on Offences and Certain Other Acts Committed on Board Aircraft. The Protocol has yet to come into force, awaiting the necessary 22 ratifications, with only 20 State Parties (as of July 2019).[235] The Montreal Protocol 2014 primarily expands the scope of jurisdiction and extradition, and clarifies the powers of the commander.[236] As a result, it does not add anything directly relevant for cybersecurity.

7.3 General Issues

The Conventions listed above have a wider scope than simply listing acts and offences. They also cover, *inter alia*, other aspects.

1. Definitions: e.g. 'in flight' or 'in service'.
2. Exclusion of military aircraft.
3. Limitations of scope to domestic and international flights.
4. Custody of alleged offender.
5. Prosecution of alleged offender.

233 ICAO, Draft Report on the Work of the Legal Committee During its 34th Session, 2-14, LC/34-WP/5-2, 16 September 2009, at. 2.98.

234 M. Smith, 2017.

235 Protocol to Amend the Convention on Offences and Certain Other Acts Committed on Board Aircraft, Montreal (Montreal Protocol 2014), 4 April 2014 (not in force).

236 M. Wukoschitz, 'ICAO: Tokyo Convention Amended by New Protocol', International Forum of Travel and Tourism Advocates', 6 May 2014, http://www.iftta.org/news/icao-tokyo-convention-amended-new-protocol.

6. Extradition of alleged offender.
7. Location of attack: e.g. on board.
8. Jurisdiction.
9. Exemption of certain motives: e.g. political, racial or religious.

As these topics are well covered in other legal texts[237] and it is argued that they do not pose any new issue within the context of cybersecurity, it is not necessary to discuss these within the scope of this work. However, this is not to say that a cyberattack would not have relevance for these areas. For example, Article 2 of the Tokyo Convention states that "no provision of this Convention shall be interpreted as authorizing or requiring any action in respect of offences against penal laws of a political nature or those based on racial or religious discrimination". Possible motives behind a cyberattack are political, racial or religious. Thus there is a clear link between Article 2 and a potential cyberattack.

As cyberattacks can allow the attacker to operate from anywhere in the world with remote capabilities, jurisdiction is very relevant. As, for example, Article 4(2) of The Hague Convention states that every "Contracting State shall likewise take such measures as may be necessary to establish its jurisdiction over the offence in the case where the alleged offender is present in its territory and it does not extradite him pursuant to Article 8 to any of the States mentioned in paragraph 1 of this Article". Thus, this could give States jurisdiction when the attacker is located in its territory, although it has no direct link to the flight.

A third example is the frequent requirement that the act or offence must be committed on board. It is only applicable to acts committed on board the aircraft, so it will not include a remote cyberattack, such as spoofing which can amount to hijacking.

These three examples depict the relevance of the Conventions to cybersecurity, however, these are just general considerations with which cyberattacks are just another subject matter.

237 *See*, P. Mendes de Leon, 2017, at Ch. 12.

7.4 Liability Conventions

The General Risks Convention[238] and the Unlawful Interference Compensation Convention[239] where both developed under the auspices of ICAO in 2009 to deal with compensation to third parties that suffer damage by an aircraft in flight while conducting an international service. Both Conventions have yet to enter into force, with 12 and 9 State Parties respectively (as of July 2019). The Unlawful Interference Compensation Convention deals specifically with compensation resulting from unlawful interference. It determines that:

> an 'act of unlawful interference' means an act which is defined as an offence in the Convention for the Suppression of Unlawful Seizure of Aircraft, Signed at The Hague on 16 December 1970, or the Convention for the Suppression of Unlawful Acts Against the Safety of Civil Aviation, Signed at Montréal on 23 September 1971, and any amendment in force at the time of the event.

As a result, there is a direct link between the liability regime and the criminal regime, whereby The Hague Convention is used to interpret the term 'act of unlawful interference'.[240]

If this instrument comes into force, then it could provide compensation to certain victims of unlawful interference. If such an act qualifies as a cyberattack, then this will be the first aviation liability treaty directly dealing with this topic. Other treaties, such as the Montreal Convention for the Unification of Certain Rules for International Carriage by Air 1999,[241] may provide compensation to passengers that are victims of a cyberattack, but it does not address this issue directly.[242]

---

238 Convention on Compensation for Damage Caused by Aircraft to Third Parties, Montreal, (Doc. 9919), 2 May 2009.

239 Convention on Compensation for Damage to Third Parties, Resulting from Acts of Unlawful Interference Involving Aircraft, Montreal, (Doc. 9920), 2 May 2009.

240 *See*, for more information, ICAO, Administrative Package for Ratification of or Accession to the Unlawful Interference Compensation Convention, https://www.icao.int/secretariat/legal/Administrative%20Packages/uicc_en.pdf.

241 Montreal Convention for the Unification of Certain Rules for International Carriage by Air, 2242 U.N.T.S. 309; S. Treaty Doc. No. 106-45 (2000).

242 *See*, B. Scott and A. Trimarchi, 2017.

7.5 *Cybersecurity Treaties (Budapest Convention)*

7.5.1 Introduction

The Convention on Cybercrime[243] (Budapest Convention) and its Explanatory Report were adopted by the Committee of Ministers of the Council of Europe during its 109th Session that took place on 8 November 2001. The Convention was opened for signature in Budapest, Hungary on 23 November 2001 and has 63 State Parties (as per July 2019) which includes all of the Council of Europe Member States – apart from Ireland, San Marino and the Russian Federation – and also includes several non-members of the Council of Europe, with notable States being Australia, Canada, Israel, Japan and the US.[244]

7.5.2 Cybersecurity Offences

The Convention has three core aims:
1. The Convention aims to harmonise national substantive criminal law elements of offences which translates to ensuring certain acts are criminalised across all Member States.
2. The Convention extends harmonisation so that national criminal law has procedural law powers necessary for the investigation and prosecution for such offences under its scope.
3. The Convention tries to set up a fast and effective regime of international co-operation in respect to cybercrime.[245]

In order to achieve its aims, the Convention is divided into four Chapters:
1. Chapter I: Use of Terms.
2. Chapter II: Measures to be Taken at the National Level.
3. Chapter III: International Co-operation.
4. Chapter IV: Final Clauses.

Chapter I simply contains four technical definitions that aim to assist with clarity in applying the Convention and are not relevant to this study.

Chapter II, Section 1 hosts nine different acts that should be criminalised by Member States at a national level.
1. Article 2 – Illegal Access.

243 Budapest Convention 2001.

244 Council of Europe, 'Parties/Observers to the Budapest Convention and Observer Organisations to the T-CY', https://www.coe.int/en/web/cybercrime/parties-observers.

245 Council of Europe, 'Explanatory Report to the Convention on Cybercrime Budapest', European Treaty Series – No. 185, 23.XI.2001, https://rm.coe.int/16800cce5b.

2. Article 3 – Illegal Interception.
3. Article 4 – Data Interference.
4. Article 5 – System Interference.
5. Article 6 – Misuse of Devices.
6. Article 7 – Computer-Related Forgery.
7. Article 8 – Computer-Related Fraud.
8. Article 9 – Offences Related to Child Pornography.
9. Article 10 – Offences Related to Infringements of Copyright and Related Rights.

The offences contained above are of general application and do not focus on a specific sector or computer system. For example, Article 2:

> Each Party shall adopt such legislative and other measures as may be necessary to establish as criminal offences under its domestic law, when committed intentionally, the access to the whole or any part of a computer system without right. A Party may require that the offence be committed by infringing security measures, with the intent of obtaining computer data or other dishonest intent, or in relation to a computer system that is connected to another computer system.

It is, therefore, clear that acts like the LOT Airlines attack[246] would fall under Article 2 as it involved access to the computers used by LOT, either by an unauthorised outsider or an insider acting *ultra vires*. Therefore, the Articles have relevance for aviation, whether this be ground systems, aircraft or the electronic links between such stakeholders.

A link between the crimes listed above and Regulation 2018/1139 can be established due to Article 122 on Combating Fraud of Regulation 2018/1139. Thus, the Regulation may have wide implications for cybersecurity as the scope of Article 122 extends beyond direct aviation activities.

In addition, it has three titles that deal with connected provisions in the area of computer or computer-related crime.

1. Article 11 – Attempt and Aiding or Abetting.
2. Article 12 – Corporate Liability.
3. Article 13 – Sanctions and Measures.

Therefore, the scope of those that can commit an offence is broadened to include not just those successfully implementing the cyberattack.

246 *See*, Chapter I.1.

7.5.3 Procedural Elements

Chapter II Section 2 contains the procedural rules, "the scope of which goes beyond the offences defined in Section 1 in that it applies to any offence committed by means of a computer system or the evidence of which is in electronic form – [in which it] determines first the common conditions and safeguards, applicable to all procedural powers in this Chapter".[247] Following this, Chapter II then contains procedural powers on expedited preservation of stored data; expedited preservation and partial disclosure of traffic data; production order; search and seizure of computer data; real-time collection of traffic data; and interception of content data.

7.5.4 Jurisdiction for Aircraft

Chapter II ends with the jurisdiction provisions. Article 22 on issues of jurisdiction requires each State Party to establish national law, where necessary, that establishes jurisdiction for the offences outlined in Articles 2-11, when committed:

a. in its territory; or
b. on board a ship flying the flag of that Party; or
c. on board an aircraft registered under the laws of that Party; or
d. by one of its nationals, if the offence is punishable under criminal law where it was committed or if the offence is committed outside the territorial jurisdiction of any State.

Article 22 then reflects international aviation law, where jurisdiction is attached to the aircraft. Therefore, this would allow the Budapest Convention to be applicable to aircraft while the aircraft is not located in its territory at the time of the commission of the crime. This, thus, ensures that a State has jurisdiction even over the high seas and that an overflown State, which may have no other connection to the flight and offence, is not tasked with exercising jurisdiction.

The scope is phrased differently than in the criminal aviation conventions. The aviation conventions, such as the Tokyo Convention, apply to offences committed "while that aircraft is in flight or on the surface of the high seas or of any other area outside the territory of any State".[248] The term 'in flight' includes "from the moment when power is applied for the purpose of take-off until the moment when the landing run ends",[249] whereas, this is not the case for the Budapest Convention as it simply covers aircraft. Thus, it covers when the aircraft is parked, taxing or in flight, or when it has crashed,

247 Council of Europe, 2001, at 19.
248 Tokyo Convention, Art. 1(2).
249 *Ibid.*

whether it is in the registering State's territory or outside, be it over the high seas or in another State.

7.5.5 Remain Points

Chapter III contains mutual assistance in two situations: (1) Where no legal basis (treaty, reciprocal legislation, etc.) exists between parties – in which case its provisions apply – and (2) where such a basis exists – in which case the existing arrangements also apply to assistance under this Convention. It also includes rules on extradition.

Chapter IV, which concludes the Convention, hosts the final clauses, which cover the standard clauses in a treaty, such as entry into force and dispute settlement.

Conclusions

1 Summary: Is the Current Regulatory Approach in the European Union Appropriate for International Civil Air Transportation?

On 7 February 2013, the EU published its cybersecurity strategy titled Cybersecurity Strategy of the European Union: An Open, Safe and Secure Cyberspace, which aims to bring about EU-wide harmonised efforts to tackle the growing and evolving cyber threat landscape. While the strategy does not refer directly to the aviation industry, it does acknowledge the importance that information and communications technology has for transport, which includes aviation.

Two years after the adoption of the EU's cybersecurity strategy, a high-level conference on the EU Cybersecurity Strategy took place on 28 May 2015. This was a significant event as it represents the first external activity of EASA with regard to aviation cybersecurity, as depicted in the Agency's cyber-timeline. From this date on, aviation cybersecurity became more visible at an EU level, with discussions and actions happening within an aviation context, as opposed to a general one.

As the EU is competent to address aviation via Article 100(2) of the TFEU, it began to address cybersecurity within this context. EASA has been given the leading role at an EU level, which was cemented by the Bucharest Declaration and confirmed by the Krakow Declaration.

The previous EASA Regulation, Regulation 216/2008, did not mention the term 'cybersecurity' anywhere in its Articles or Annexes. However, despite the lack of clearly defined mandate for cybersecurity, this did not mean that Regulation 216/2008 was silent nor did it mean that the Agency did not have responsibilities in this area. There are several examples that could be taken from Regulation 216/2008 that provide a link between cybersecurity and the Articles and Annexes of the Regulation. As a result, although the scope of Regulation 216/2008 is limited to safety and environment, this legal regime was applicable to cybersecurity in as much as it relates to these two areas. It is argued in this study that this is not an appropriate way to regulate cybersecurity. The safety-based rules must address cybersecurity directly to ensure that this is fully covered.

Further to the Articles and Annexes of Regulation 216/2008, EASA has initiated its rulemaking procedure, starting in 2016, to include six different rulemaking tasks. While this involves legal initiatives, notably through the creation of a cybersecurity-based Regulation, the scope is much wider. It included planning and setting targets via the Roadmap, the creation of a cybersecurity strategy, stakeholder cooperation through the creation of a CERT, and finally there is the promotion of information sharing and cooperation through the creation of a common aeronautical vulnerabilities database. The steps taken by EASA are very commendable and important. If done correctly, creating new aviation safety rules or amending existing ones with cybersecurity in mind will provide for better cybersecurity across the EU. The steps also look further than regulation as it covers stakeholder cooperation and information sharing.

The area of information sharing, notably the creation of the ECCSA, shows that the EU aviation safety regime is not the only area of law that is relevant for aviation cybersecurity. This is because the NIS Directive covers the creation of CERTs within the scope of aviation. Further study shows that other areas of EU law may also be relevant, such as the GDPR, and the proposal for a regulation on ENISA and on ICT cybersecurity certification. Therefore, the wider scope must be considered.

In 2015, the Commission published its Aviation Strategy to Enhance the Competitiveness of the EU Aviation Sector. The strategy stated that the EU wants to repeal and replace Regulation 216/2008. Regulation 2018/1139 represents the first time that the term 'cybersecurity' features directly in the EASA cover Regulation. As a result, this codifies the link between aviation safety and cybersecurity within a binding legal context. However, Preamble Recital 12 makes it explicitly clear that the two are distinct areas; cybersecurity is not *stricto sensu* a part of safety. Recital 59 takes the classification of 'cybersecurity' further than Recital 12 as it places 'cybersecurity' directly under the scope of aviation 'security'. As a result, it points to EU aviation security under Regulation 300/2008 as the legal mechanism to tackle aviation cybersecurity, rather than the safety regime.

The EU has been active since 2001 in the area of aviation security. These efforts, due to the factual and international regulatory context, have focused on unlawful interference stemming from physical sources. Therefore, Regulation 300/2008 has a strong focus on preventing sabotage with explosives and *in situ* hijacking with weapons. As a result, the direct cybersecurity aspect is missing. While the current regulatory environment is an improvement on pre-9/11 times, more certainly needs to be done if cybersecurity is to be dealt with under the Regulation 300/2008 regime.

Attempts were made to find indirect links to cybersecurity, however, nothing concrete could be found within the wording of Regulation 300/2008. This does not mean that the Regulation is irrelevant for cybersecurity as Member States could take measures as the Regulation provides a level of discretion to the Member States. For example, Member

States could require cybersecurity to be a part of staff training or they could implement more stringent measures so they also include cybersecurity. While flexibility, that still maintains the set EU security standard, is useful, it still requires Member States to take the initiative, which can result in cybersecurity being unregulated at an EU level or an EU-wide fragmented aviation cybersecurity regime.

If cybersecurity were to be explicitly included within this regulatory regime, such as through the expansion of Annex I, then there appears to be appropriate tools in place to ensure implementation of the measures and oversight.

Due to the limitations of Regulation 300/2008, it can arguably be concluded that it has a limited effect on cybersecurity under the current setting. The EU may desire to bring cyberattacks under the unlawful interference regime, but as identified in Chapter III.4, this has significant drawbacks. In order to remedy this, the EU may desire to tackle cybersecurity in a *lex specialis* Regulation. However, more needs to be done to regulate cyberattacks taking into account the difference found between the aviation safety and security regimes.

It became essential to examine the Chicago Convention and its Annexes, as they are explicitly cited in Regulation 216/2008 and Regulation 2018/1139, as well as in Regulation 300/2008, which were extensively discussed in Chapter II.

It was shown that the Chicago Convention was not drafted to directly tackle cybersecurity. However, despite this, the discussions on Article 3 *bis*, 4, 5 and 6 have shown that the Chicago Convention has relevance for cybersecurity with interpretative methods. This point is extended as many of the Articles to the Chicago Convention shall be read together with the 19 Annexes.

Amendment 12 to Annex 17 includes the term 'cyber threats' which marks the first direct reference to the term 'cyber' in the Annexes. The applicability of the Annexes is not limited to Amendment 12 as, for example, the Annexes also utilise several words with potential relevance, which includes: 'authorised', 'accuracy', 'integrity', 'availability' and 'functionality'. It is, therefore, possible to read cybersecurity into SARPs even though they were not drafted to specifically tackle such an issue. Despite all of this, there are significant gaps and weaknesses in the Annexes and more needs to be done by ICAO to address all aspects of cybersecurity that are relevant for aviation.

In addition to the Annexes, ICAO has also been active in cybersecurity in other areas. For example, the ICAO 39th Assembly introduced the first Assembly Resolution focused on cybersecurity which further promotes a consistent and coherent approach in managing cyber threats, and risks on how to better address cybersecurity in civil aviation. This work is likely to be continued with the 40th Assembly. An additional example includes the Dubai Declaration which makes six specific declarations focused on the role of States in cybersecurity, collaboration and exchange of information, the role of all other stakeholders, peaceful purposes, promotion of the criminal treaties, and makes certain acts offences. This draws parallels with EU efforts.

A specific reference is made to the international criminal aviation law treaties in Regulation 300/2008 due to their relevance in dealing with prohibited activities against international civil aviation. These Conventions are also useful tools for the interpretation of Article 3 *bis*, the scope of security-based articles in ASA and for placing Annex 17 within its wider international context. It was found that they have varying relevance for cybersecurity. For example, the Tokyo Convention 1969 leaves it to domestic penal law to determine what constitutes an offence, of which States could criminalise cyberattacks. Whereas, The Hague Convention 1970, on the other hand, may criminalise cyberattacks due to its general prohibition on the use of force against aircraft. Finally, the Beijing Protocol directly addresses cyberattacks as it, *inter alia*, addresses the jamming of aircraft. International criminal air law has been supplemented with the Budapest Convention which provides that aircraft have the jurisdiction of the State of registry for crimes covered by the Budapest Convention.

There are several aviation organisations across the world that aim to promote aviation safety and security. However, the general treatment of cybersecurity under the domain of security by such organisations has resulted in the topic not being discussed in many regional organisations as their scope is often on core aviation safety aspects, such as airworthiness, maintenance and licencing. However, as cybersecurity is relevant for these areas and with potential developments of Annexes, it can be reasonably expected that the role of regional organisations will increase in importance. Despite the lack of action by regional organisations, other key stakeholders are active in research, information sharing, training and shaping policy.

As a result, Chapters II and III show that there are indirect and direct laws regulating aviation cybersecurity. However, these are fragmented and leave lacunas. Further, the dualist regulatory approach taken in aviation, that of safety and security, results in different interpretation, application and enforcement of any cyber-based rules. Therefore, the current regulatory approach in the EU is not appropriate for international civil air transportation.

2 NEXT STEPS

While the current steps taken by the EU, as well as those at an international level, provide some guidance as to the next steps, in many cases only predictions can be made. It is predicted that the next steps will take place in five core areas.

First, Regulation 2018/1139 is new, so acts by EASA pursuant to Article 88 can be expected. This will be in three areas.

1. Cooperation between Member States, the Commission and the Agency on cybersecurity, where interdependencies between civil aviation safety and security exist (Art. 88(1)).
2. The Agency shall, upon request, provide technical assistance to the Commission, where the Agency has the relevant safety expertise, in the implementation of Regulation 300/2008 (Art. 88(2)).
3. The Agency can react so to contribute to protecting civil aviation against acts of unlawful interference (Art. 88(3)).

The extent and success of these will be revealed in time via implementation.

Second, as RMT.0720 shows, there are plans to develop a specific EU aviation cybersecurity Regulation within the EU aviation safety regime. In addition, RMT.0648 sets out to amend relevant CS. As a result, it can be expected that there will be an increased number of *lex specialis* regulations and supporting documents for aviation cybersecurity under the EU aviation safety regime.

Third, the growth in EU aviation rules to combat cybersecurity are unlikely to remain within the safety regime. As there is a tendency, such as within the ICAO context, to place cybersecurity under the domain of security, with amendments taking place to Annex 17 specifically on this topic, it is likely, due to the direct link, that EU security regime under Regulation 300/2008 will be subjected to amendments to address cyber threats and vulnerabilities.

Fourth, with the increased activities of ICAO on the topic, such as with Amendment 12 to Annex 17, it can be expected that more amendments to Annexes will come. It is less likely that ICAO will facilitate a treaty to deal more directly with cybersecurity due to the relative lack of interest from States in adopting new criminal air law treaties. However, this may change with the coming into force of the Beijing Convention and Protocol which has direct application to cybersecurity.

Finally, it can be expected that non-legal actions will increase. For example, there have been many initiatives for cooperation and information sharing, with discussions also focusing on the development of industry standards. It appears that stakeholders see the relevance of these actions so it can be expected that such actions will take place. This is increased by the growing threat of cyberattacks, so stakeholders are investing in all of the four control types.

Cybersecurity is important for all aviation stakeholders as numerous and varying threats and vulnerabilities are present. This topic will continue to grow in importance and much more needs to be done in order to provide an acceptable level of safety and security for aviation in the EU and globally. Regulation is only one part of this, but an important part where the efforts of those involved are going in the right direction. However, more needs to be done, and more appreciation of the multifaceted nature of cybersecurity and cyber-attacks needs to be implemented into the rules so all areas are covered.

3 Cybersecurity Considerations for Aviation Regulation

3.1 *Introductory Remarks*

For regulators to successfully combat cyberattacks and promote cybersecurity in aviation, there are overarching cybersecurity factors that must be taken into consideration when drafting and then applying the rules. This section of the conclusion has collected generic points, found in other sectors that must be considered by aviation stakeholders when regulating cybersecurity.

3.2 *Control Types*

Control types refer to the overarching methods of defending against a cyberattack.[1] Four different types of controls can be identified: Physical, technical, procedural and legal.

3.2.1 Physical Controls

Physical controls refer to "measures designed to deter, prevent, detect or alert unauthorized real-world access to a site or material item".[2] In other words, this is physical security

1 Raef Meeuwisse, *Cybersecurity for Beginners*, First Edition, (Lulu Publishing Services, 2015), at 48-50. *See*, for an example of an industry information document on increasing control types, Verizon, 2018 Data Breach Investigations Report, https://www.verizonenterprise.com/resources/reports/rp_DBIR_2018_Report_execsummary_en_xg.pdf.

2 R. Meeuwisse, 2015, at 49.

designed to stop acts like unauthorised entry into a building or the use of security cabinets to prevent ease of access to confidential data stored on a memory card. Physical security is often overlooked in the context of cybersecurity as it does not necessarily require a computer. However, without such controls, sophisticated cyberattacks become unnecessary as the attacker could simply walk into a building and walk out with sensitive data with little investment in time, money and other resources. This would then negate any technical controls that have been put in place so resulting in them being a wasted investment by the victim.

3.2.2 Technical Controls

Technical controls are often the focus for cybersecurity and include, *inter alia*, anti-malware software, firewalls and encryption.

> Firewall: A network security system that monitors incoming and outgoing network traffic and controls it based on predetermined security rules.
>
> Antivirus: Software that detects and destroys viruses.
>
> Encryption: The process of encoding information so that only authorised parties can have access to it.

These are "the use of an electronic or digital method to influence or command how something is or is not able to be used".[3]

3.2.3 Procedural Controls

Procedural controls are "the use of a sequence of steps to influence or command how something is or is not able to be used".[4] This is the establishment of rules, working methods or steps in which authorised users should operate or the setting of controls in the technology to ensure it is used in an authorised way. One such control can be the prohibition of the use of external USB sticks as these can introduce malware from less secure computers. One method used by hackers is to drop USB sticks outside the victim's property in the hope that an employee finds the infected USB and then insert it into a computer. Here, training is of key importance.

---

3 *Ibid.*, at 50.

4 *Ibid.*

3.2.4 Legal Controls

The final stage is legal controls. This is "the use of legislation to help promote and invest in positive security methods and also to deter and correct infringements".[5] This, therefore, has two concerns.

First, the inclusion of effective legislation providing entities with a cybersecurity standard that they should reach, rather than allowing market forces to dictate this alone. This can be seen with data protection laws, which often contain specifications on the storage of data so to reduce the chance of it being stolen. Thus, it is proactive.

Second, legal controls can dissuade certain behaviour. For example, they can criminalise cyberattacks to allow authorities to prosecute attackers, which also can act as a deterrent to potential future attackers if they know that they could face repercussions, such as heavy fines and imprisonment. It is also important that both the aviation safety and security regimes act within this as they each have a different focus and require different responses.

3.2.5 Putting It Together

In order to have effective cybersecurity, all four control types need to be present and effective. For example, if there are insufficient technical controls, then a victim may never know that they have been attacked or the lack of tools may not allow the attacker to be identified. If this is the case, then the need for legal controls is null and void. This is because legal sanctions cannot be placed on an attacker when identification is impossible. This study has focused mainly on legal control types within aviation as its aim is directed towards current and future regulations.

3.3 Control Modes

The control types have to take into consideration the different control modes. The term 'control modes' is an umbrella term used to incorporate preventative, detective and corrective defence methods. Each one represents a different time posture.

3.3.1 Preventative

Preventative refers to controls that are designed to stop the attack before it is successful. For example, prevention control modes may be strengthened by a firewall, pen-testing, extensive staff training and an advanced internal cybersecurity policy.

---

5 *Ibid.*

3.3.2 Detective

Detective refers to controls that are designed to monitor and alert the victim to an attack or potential attack. One method of achieving this is through real-time antivirus software that is constantly running in the background.

3.3.3 Corrective

Corrective refers to controls that are designed to rectify the consequences of the attack or at least mitigate the damage to the greatest extent possible. Measures that focus on corrective action can include having a critical response procedure, such as a CERT, in place that is initiated once an attack has been discovered.

3.3.4 Putting It Together

As these three cover the whole event, from pre-attack to post-attack, all modes need to be guaranteed, through all control types, in order to help protect against cyberattacks.

3.4 *Life Cycle of a Cyberattack*

A cyberattack is not one act that takes place at one moment. Rather, it comprises several different steps tailored for the specific victim that will serve the ultimate goal of the attacker. It is argued that it is crucial to appreciate the evolution of a specific cyberattack as this will allow victims to better fight against such invasions and for regulators to produce compressive rules that govern all stages of the attack.

The steps highlighted below may be present in any cyberattack, however, each attack is unique. As a result, not all steps may be present, the steps may be carried out in a different order to that which is presented here, or the focus and duration of one step may be quite different from another. Therefore, the specifics of each cyberattack should be appreciated when assessing these steps. This section of the conclusion will, however, look at a cyberattack generally in order to be as inclusive as possible and, therefore, covering as many steps as possible.

3.4.1 Initial Reconnaissance

In order to launch a cyberattack against a victim, some sort of initial reconnaissance is required to know who, what and how to attack, and what the consequences would be of such an attack. Reconnaissance may be as simple as confirming that the victim has a public access website, so a DDoS attack can be launched. It could also require more research, for example, information gathering from "social networking websites, Internet search engines, conferences, lectures, academic material, underground repositories",[6] to provide the required intelligence to profile the targeted individual or organisation. It can also include more intrusive measures (active reconnaissance). This could involve standing outside the victim's premises, following employees or stealing equipment.

In aviation, this can take many forms due to the broad range of stakeholders and access that the public has to aircraft and airports. It could then involve visiting airports or taking flights to observe weaknesses. Alternatively, due to the advanced safety culture in aviation, it could also involve acquiring maintenance manuals that are often available, sometimes for free, to the public and studying them.

Passive reconnaissance may also be one way of gaining essential information. They "consist of analysing traffic, decrypting the unsophisticated traffic, monitoring weakly protected communications, and intercepting authentication information and credentials such as passwords".[7] Thus, it involves observing what the victim is doing in the cyber world, rather than actively engaging with the victim. This can then provide the attacker with essential information to find vulnerabilities that allow for the intended cyberattack to take place.

Depending on the methods adopted by the attacker, some may be observable by taking basic precautions. For example, a large company is likely to have some sort of basic physical security at its offices and this could deal with unauthorised access and individuals spending prolonged time outside the building. However, others are less easy to tackle, especially passive-based reconnaissance. Therefore, basic measures cannot guarantee the prevention of an attack, but the more precautions that are put in place, the harder it is for the attacker to gain essential information, which may result in the attacker being caught before they implement the attack.

3.4.2 Penetration and Gaining a Foothold

Penetration is the step where the attacker compromises the victim's device. As highlighted in Chapter I.1.2, this can take place via many methods, such as malware and phishing. Different attack methods will penetrate to different levels. A worm can penetrate much deeper than, for example, a bomb threat via social media, which is superficial.

6 A. Smith and N. Toppel, 'Case Study: Using Security Awareness to Combat the Advanced Persistent Threat', Paper Presented at the 13th Colloquium for Information Systems Security Education (CISSE), Seattle, 1-3 June 2009.

7 Infosec Institute, 'The Cyber Exploitation Life Cycle', 22 March 2013, http://www.resources.infosecinstitute.com/the-cyber-exploitation-life-cycle/#gref.

Once access has been gained, it is important for the attacker to obtain a foothold so as to ensure that they have the required access and control of the technology. This can include creating a backdoor in the software, and then creating a command and control infrastructure. This will then allow the attacker to enter the computer and then control it in such a way so as to fulfil the intended results of the attack. Another is by appropriating privileges. This can be done by obtaining the victim's username and password, so that access to the system can be gained without having to crack the password.

3.4.3 Following Access

Once access is gained, the attacker can conduct internal exploration. This is where information is collected on the system's internal environment. This can be done to allow future attacks to be more successful and carried out with ease as it may highlight previously unknown vulnerabilities; it may allow the attacker to navigate the system more easily as there will be a greater understanding of it; and it can divulge the location of sensitive files and documents, databases, or the contents of email accounts which are stashed in files or email servers and domain controllers.

The attacker may, once in a system, extend the attack to other computers within the same network. This is called 'lateral movement'. Access to these computers may not have been possible prior to the attack, however, through appropriating privileges, such as passwords, there may now be access to these computers and the hosted data.

Once the attacker is in, there is a requirement to remain there so the intended goal can be achieved, rather than being spotted or having control taken away. This can be achieved via three methods.

1. The installation of additional malware: Once access is gained, including more malware becomes much simpler and this can allow the attacker to maintain presence. Furthermore, if any malware is detected and deleted, the inclusion of additional malware secures the maintained presence.
2. The use of legitimate credentials: The attacker can be disguised as a legitimate user, such as by using a valid VPN.
3. The use of log-in details: The attacker can steal log-in details from the victim and then use them to gain unauthorised access via genuine means.

This will then allow the attacker to continue to work within the computer and, due to the clandestine nature, remain undetected.

3.4.4 Exfiltration

The final stage is transferring the extracted data from the victim's computers to the attacker or selected third parties. Depending on the type of extracted data, this can have significant financial and reputational damages for the victim. However, it may also be the case that no data is extracted and the purpose of the cyberattack was to destroy the system, make it inaccessible as with ransomware or a DDoS attack, or to simply prove a point.

3.4.5 Putting It Together

Appreciating the life cycle of an attack is crucial for cybersecurity and its regulation. This is because, if a cyberattack is viewed as one single event, such as the extraction or the exploitation of confidential information, this then ignores all the prior steps that went into the exploitation. This could lead to a focus being on retrieving the stolen data and prosecuting the attackers. However, as shown in Chapter I.4.9 the attacker is often not found so it may be a waste of resources only pursuing the attacker. Whereas, it may be more suitable to focus on preventative measures during the initial reconnaissance stage, including software to seek out on-going attacks, and strengthen attacked systems so any internal reconnaissance data on vulnerabilities is no longer valid. In total, the average preparation stage is five months and the average executive stage is one hour, whereby the worldwide average for detection of a cyber attach is around 100 days.

3.5 Digital Defence Points

An attack method may target different points of the victim which need to be protected. This is called the 'digital defence point' and can be categorised into one of five distinct subjects.

1. Data: That which is stored in electronic or digital format.
2. Devices: Any hardware, such as computers and smart phones, used to create, modify, process, store or transmit data.
3. Applications: This is any software on a computer. It is usually a programme that exists to create, modify, process, store, inspect or transmit specific types of data.
4. Systems: Groups of applications that operate together to serve a more complex purpose.
5. Networks: A collection of devices, wiring and applications used to connect, carry, broadcast, monitor or safeguard data.

The attack strategy can focus on any of the above five locations depending on which the victim possesses. If the attacker wishes to cause maximum disruption to the victim then the network may be a more suitable attack location, whereas if the attacker wishes to extract sensitive data, then the data itself is likely to be the direct target. The five are connected, as a system is a group of applications and applications require host devices, and data may be on those devices. Therefore, it is likely that an attack will use more than one digital defence point in an attack.

Identifying these five allows potential victims to put into place appropriate cybersecurity measures customised to the specific vulnerabilities of each digital defence point, as opposed to having generic measures that are not tailored for the specifics of each point.

3.6 Common Goals

Cybersecurity measures should not simply prevent attackers from limiting the use of a technology or deny the acquisition of certain data or information. There should be three levels to any cybersecurity measure. When regulating data and information, cybersecurity regulators must consider the CIA Triad which is an acronym for 'confidentiality', 'integrity' and 'availability'.[8]

3.6.1 Confidentiality

Confidentiality refers to the denial of access of unauthorised entities to certain information. In other words, it means keeping data and information private. As data and information can be valuable and its acquisition is the goal of some cyberattacks, this is important to achieve. This can include sensitive data, such as trade secrets, confidential information including national secrets, and intellectual property, which can have certain national, regional and international legal protection. Confidentiality is supported by technical tools which can include password secure access and encryption.

3.6.2 Integrity

Integrity is ensuring that systems, information and data are accurate and remain unchanged by unauthorised entities. This is often the most difficult to spot. This can be demonstrated via the infamous Stuxnet attack which affected Iranian nuclear centrifuges that resulted in a sabotage, but told the user that they were functioning correctly.

3.6.3 Availability

Availability is ensuring that the system can be used or the information accessed by authorised users when desired. If authorised persons cannot access and use the hosted information, then its utility is lost and the rationale for hosting it in the first place is questionable. Thus, it is important that it can be used as intended.

8 P. Singer and A. Friedman, 2014, at 35.

3.6.4 Putting It Together

The aim of cybersecurity is not simply denying access to unauthorised users, but also extends to controlling the system and its hosted data. Regulators should ensure that these three are achieved and maintained at all times.

4 Aviation Considerations for Cybersecurity Regulation

4.1 Introductory Remarks

For regulators to successfully combat cyberattacks and promote cybersecurity in aviation, there are overarching aviation factors that must be taken into consideration when drafting and then applying the rules. This section of the conclusion has collected generic points, found in other sectors that must be considered by aviation stakeholders when regulating cybersecurity. By analysing all the discussions on cybersecurity by the relevant stakeholders, both in aviation and non-aviation, the following activities can be identified.

1. Regulations.
2. Development of industry standards.
3. Information sharing and reporting.
4. Risk assessments.
5. Training.
6. Commitment of resources.
7. Awareness Promotion.

It is argued that all of the below should be fully covered by the EU cybersecurity. However, as analysed through Chapters II and III, not all of these are fully covered.

4.2 Regulations

Regulations are only now beginning to appear that directly tackle aviation cybersecurity. As a result, there are significant gaps in the regulatory system. However, EASA is taking steps in its Rulemaking Programme to address this as it plans to draft a *lex specialis* Commission Regulation focused on cybersecurity and support the certification system by amending CS to address cybersecurity.

The limits of EASA's competencies must be appreciated. As a result, the efforts discussed in Chapter II do not cover every legal domain that could be relevant for cybersecurity. However, this limitation is correctly appreciated by EASA and it acknowledges the role of other legal instruments, such as the NIS Directive and GDPR.

The EU aviation legal regime also correctly appreciates its limits. As a result, the EU has correctly referred to international law, notably the Chicago Convention, which is binding on the EU Member States. However, reliance on the Chicago Convention, its Annexes and the criminal air law treaties has produced gaps in the EU regulatory system due to ICAO's focus on 'unlawful interference'.

The development of new rules needs to be based on a new approach, one that breaks away from the dichotomous safety v. security way of thinking that has dominated so far. There is a tendency to classify cybersecurity as a part of security. As a result, safety-based regulations tend to ignore this issue, and instead leave it to the security-based rules. This must be stopped as a cyberattack can have a grave impact on safety. Either cybersecurity should be inserted into the full aviation legal regime, such as amending Regulation 300/2008 and Regulation 2018/1139, to properly address this issue, or a separate legal instrument or separate instruments should be adopted to directly deal with cybersecurity in aviation.

4.3 Development of Industry Standards

The efforts made by the EU also show that not everything can or should be done through regulations. It acknowledges the importance of developing industry standards which are produced by those with primary knowledge and can be flexible to suit the needs of the user. Thus, there is a necessity for coordination, information sharing, event investigation and response.

Both the Bucharest Declaration and the Krakow Declaration promote the use of industry standards for aviation cybersecurity. This is supported by the development of the Strategy for Cybersecurity in Aviation. As set out in SPT.071, the strategy will include an action for the development of industry standards. Finally, this is codified in Regulation 2018/1139 as Recital 12 of the Preamble states that "[u]se should be made of recognised industry standards and practices, where it has been found that they ensure compliance with the essential requirements set out in this Regulation".

This conforms to a wider trend for the use of industry standards as it can be seen in other areas of aviation regulation, such as with unmanned aircraft. The use of industry standards in EU aviation regulation is increasing as it allows EASA to utilise industry expertise, and provides for a level of flexibility and evolution as they do not have to go through the slow EU rulemaking system. However, compliance to these standards is not obliged in law and they have yet to be developed for aviation, so the success of this will unfold in the future.

4.4 *Information Sharing and Reporting*

Cooperation between stakeholders is seen as important. This can be done generally through existing mechanisms, which includes the mandatory reporting system under Regulation 996/2010 or via voluntary reporting based on trust, such as EASA's Confidential Safety Reporting system. It can also be done more specifically within the context of a cybersecurity forum, for example, through the ECCSA. Such activities should be encouraged as, for example, it can facilitate the reduction of vulnerabilities across the sector. This should be encouraged, but challenges are present, such as under reporting, exclusion of certain stakeholders or fragmented sharing groups.

4.5 *Risk Assessments*

A part of cybersecurity knowledge is knowing the present risks and vulnerabilities. Therefore, stakeholders should be active in risk assessments that are coherent, comparable, share common risk criteria, and share terminologies and methodologies. Risk assessments can be found in several sources of law, such as the NIS Directive and Regulation 300/2008, as well as in Annex 17. Stakeholders are also active in this area. For example, CANSO's information sharing role has included cybersecurity and in June 2014, it published its Cyber Security and Risk Assessment Guide. Here, clarification may be required to standardise and harmonise all of these efforts.

4.6 *Training*

Currently the EU aviation rules are silent on cyber-based training. However, both the EU aviation safety and security regimes cover the training of certain personnel, such as flight and cabin crew, and ground staff. With amendments to existing regulations or the creation of the new aviation cybersecurity regulation, it is probable training will be mandated to deal with this area. However, until this happens, training is left to other stakeholders, like IATA's non-mandated cybersecurity classroom. This gap needs to be addressed to ensure that vulnerabilities are mitigated.

4.7 *Commitment of Resources*

Layer 7 of the Bucharest Declaration declares that all stakeholders should allocate resources, whether human, monetary or technical as appropriate for aviation cybersecurity. The inclusive nature of this allows for all aspects to be covered under the heading 'resources' so it should not leave any gaps. However, the subjective nature of this layer, provided through the wording 'appropriate level', could result in a disproportionate allocation of resources. More needs to be done to quantify the expected minimal allocation of resources, to ensure that allocation has been implemented and to support those that cannot meet the set requirements.

4.8 *Awareness Promotion*

There is a lot of focus on structured information sharing platforms, such as the ECCSA, which facilitates the dissemination of information about aviation cybersecurity. However, as cybersecurity affects all stakeholders, from airlines and manufactures to passengers, information should be appropriately shared to all. A lesson can be taken from unmanned aircraft. These aircraft are flown by a diverse demographic, so easy to understand information leaflets and videos have been produced by EASA, as well as proposed technical rules in the form of a Commission Regulation, so all stakeholders have relatable information to help in understanding and applying the applicable laws. This could be done for cybersecurity to educate stakeholders, such as passengers, about the risks of interfering with on board WiFi or the use of jammers near airports. In other words, efforts are needed to promote the rules and consequences of violating them to all stakeholders and not just those heavily involved in aviation and cybersecurity.

5 Six Basic Assertions

5.1 *Cybersecurity Will Grow in Importance for Civil Aviation*

When analysing the number of publicly reported attacks as identified in Table 1, there is a growing trend in the number of cyberattacks against aviation stakeholders with only one taking place in the 1990s, four taking place in the 2000s and 45 so far taking place in 2010s. There is, thus, a notable increase between the decades.

The limitations of the study must be appreciated. First, the increase is only a general trend as 2017, the last complete year for this investigation, only saw one publicly recorded cyberattack. Until more information is received on later years, it cannot be assessed whether this is a statistical outlier or the start of a new trend.

Second, Table 1 is a non-exhaustive list as it only documents publicly recorded cyberattacks against aviation stakeholders that could be found. It is possible and very likely that many such cyberattacks either go unreported, the victims remain unaware or stay out of the public domain.

Despite the limitations, when compiling the publicly reported attacks, from their fragmented sources, it is evident that there has been a rise in frequency since the first cyberattack in 1997.

It is argued that this trend will continue for a number of reasons:

1. All aviation stakeholders are increasingly using and becoming reliant on technology, such as computers.
2. The computers are becoming more connected, such as with the IoT, communication and navigation.
3. Automation in aviation is prevalent and a component in all of the main activities of aviation stakeholders.
4. Aviation provides different and numerous motives to different potential attackers, so it presents itself as a target. This could be due to a stakeholder having a lot of capital, holding a lot of personal passenger data, being exposed very publicly, or causing stress to some passengers that want revenge.
5. Cyberattacks are becoming the preferred tool for some individuals and groups over more traditional physical methods.

Therefore, what can be concluded is that aviation is increasingly relying on computers in all aspects. These computers are increasingly being connected to the Internet and other networks, as well as becoming more automatic. Therefore, the aviation sector is producing more vulnerabilities that an attacker can exploit. In addition, the aviation sector is an appealing target, so there are many reasons why aviation is being subjected to a growing number of cyberattacks. None of these are likely to cease, so this problem is unlikely to fade away.

These findings and conclusions are not unique to aviation. For example, other transportation sectors, like space, road, rail and water, are becoming more reliant on computers, the IoT and automation. This is also the case for other sectors like finance and energy. Therefore, while the growth in numbers of cyberattacks can be explained by an analysis of the aviation sector, the causes are not confined to this sector and quite simply, aviation is just another target. However, the consequences may be quite different. Due to the overlap in sectors, this is justification for cross-sectorial cooperation in order promote cybersecurity.

5.2 *Cybersecurity Is Not One Homogenised Topic, But Is an Overarching Title That Encompasses Numerous Acts, Different Actors, Varying Victims and a Plethora of Consequences*

This work has also provided strong evidence that cybersecurity should be considered as an overarching title. This is because cyberattacks against civil aviation have:

1. utilised numerous attack methods, such as DDoS attacks, jamming, malware, unauthorised access;
2. been implemented by different individuals and groups, such as teenagers with limited resources, nation States, activist groups and by anonymous sources;
3. been targeted towards different stakeholder, such as airlines, airports, air traffic control, law firms and authorities; and
4. been motivated by different reasons and produced varying consequences, such as to make a political statement, steal information or money, or cause disruption.

More than one stakeholder may be affected by a single attack, other attack methods may be utilised, different attackers may enter the scene with different motives, and this may produce new consequences. Therefore, not only is the term 'cybersecurity' an overarching title, it is also an evolving one.

5.3 *Regulation Should Not Be the Only Tool in Promoting Cybersecurity*

In order to have effective cybersecurity, all four control types need to be present and effective:

1. Physical Controls.
2. Technical Controls.
3. Procedural Controls.
4. Legal Controls.

As argued in Conclusion 2.2. all four controls need to be in place to have effective cybersecurity. Regulations only account for one element of these different controls.

Legal controls can take two forms, either proactive or reactive. First, the inclusion of effective legislation providing entities with a cybersecurity standard that they should reach, rather than allowing market forces to dictate this alone. Second, legal controls can be reactive. For example, they can criminalise cyberattacks to allow authorities to prosecute attackers, which also can act as a deterrent to potential future attackers if they know that they could face repercussions, such as heavy fines and imprisonment. As a result, regulating cybersecurity is not a one size fits all approach, it is multifaceted.

Legal controls can also be used to implement or strengthen the other controls. For example, the law can mandate certain physical security (physical controls), the use of cybersecurity software (technical controls) or employee training (procedural controls). Therefore, the regulation of cybersecurity has an additional layer to which must be considered.

Despite the above point, the other controls can exist independently from legal controls, as they can be implemented via industry standards or simply because they make commercial sense.

It is, therefore, argued that regulations cannot be the only tool used in cybersecurity by aviation stakeholders. This conclusion appears to be generally shared, as the actions of EASA, as well as other stakeholders, also involve cooperation, information sharing, the provision of training, the production of industry standards and have taken cybersecurity steps without being legally mandated.

5.4 *Role of the European Union in Aviation Cybersecurity Will Increase*

Aviation cybersecurity in the EU did not really begin to be discussed until 2015. As a result, its explicit reference from key EU Regulations – Regulation 216/2008 and Regulation 300/2008 – is missing.

In 2015, the Commission published its Aviation Strategy to Enhance the Competitiveness of the EU Aviation Sector where the strategy stated that the EU wants to repeal and replace Regulation 216/2008. In the 2015 Proposal, the term 'cybersecurity' featured in the text and was included as a political statement to make it explicitly clear that the Regulation includes 'cybersecurity' under the heading 'security'.

From then, several steps have been implemented by EASA in its Rulemaking Programme to regulate and discuss cybersecurity:

1. RMT.0648 – Aircraft cybersecurity.
2. SPT.071 – Cybersecurity road map.
3. SPT.072 – Aviation computer emergency response team (AV-CERT).
4. RMT.0720 – Cybersecurity risks.
5. SPT.071 – Strategy for Cybersecurity in Aviation.
6. RES.012 – Cybersecurity: common aeronautical vulnerabilities database.

There is a clear trend that the EU, with EASA being the central focus, will take a lead in aviation cybersecurity. This is confirmed by the declarations made in the Bucharest Declaration and the Krakow Declaration.

The EU's increased involvement should not be seen as limited to aviation specifically. The increased attention that aviation stakeholders have given to cybersecurity falls within the wider context of the EU's cybersecurity and digital single market strategies. Efforts in this area also affect aviation stakeholders, such as the proposal for a regulation on ENISA and on ICT cybersecurity certification. Therefore, the EU's involvement is wider and will likely continue to grow.

5.5 *Regulators Must Work More Closely with Stakeholders from All Domains and Not Just Aviation Stakeholders or Member States*

It has been shown that all aviation stakeholders are vulnerable to cyberattacks. As a result, all aviation stakeholders must have appropriate cybersecurity to cover vulnerabilities and mitigate risks. As a part of this will involve regulations, regulators must work with these stakeholders in order to produce regulations that are suitable for the nuances of each stakeholder and this is what has taken place through extensive stakeholder consolation by EASA.

The Commission, via DG-MOVE and EASA, has appreciated that it is not able to produce cybersecurity regulations and non-legal actions without the support of other public actors. Therefore, it has drawn on the expertise of ENISA, EU-CERT, other EU Institutions and Member States.

Finally, as seen with the Bucharest Declaration and Krakow Declaration, cybersecurity stakeholders have also been invited to discussions. It is, therefore, evident that the EU sees the relevance of non-aviation stakeholders.

Within the context of EU aviation cybersecurity, more could be done to introduce external aviation and non-aviation stakeholders. For example, non-EU manufactures could be invited to participate in information sharing. This would allow worldwide expertise to influence cybersecurity in the EU.

5.6 *Convention on International Civil Aviation and Its 19 Annexes Have Not Gone Far Enough to Effectively and Efficiently Regulate Aviation Cybersecurity*

As shown in the conclusions found above in this chapter, the Chicago Convention and its Annexes are relevant for EU aviation. However, direct relevance is extremely limited, with the only reference being found in Amendment 12 to Annex 17. Indirect reference, such as via ASA or through interpreting safety-based SARPs to include cybersecurity do provide for more coverage. However, this is fragmented and provides great legal uncertainty.

The above point is compounded by ICAO's insistent treatment of cybersecurity as an act of unlawful interference. This is extremely narrow as not all cyberattacks are unlawful or cause interference. If this approach is maintained, then it is unlikely that the international civil aviation regime will evolve to appropriately regulate cybersecurity and combat cyberattacks. More, therefore, needs to be done with a wider approach to include both safety and security, and even financial interests.

As with the EU, ICAO's response to cybersecurity is quite young and more output can be expected in the future.

6 Overall Conclusion

In conclusion, while the regulation of cybersecurity at an EU level may *prima facia* appear very minimal, this is certainly not the case. Despite this, the efforts are by no means complete and are very much in the developmental stage. While the current efforts should be commended, the efforts – as outlined in this work – must be improved so the risks are better understood and managed, cyberattacks are properly managed and responded to, and cybersecurity is correctly promoted.

About the Author

Dr Benjamyn I. Scott graduated in 2011 in Law at the University of Kent and the University of Amsterdam (LL.B. (Hons)). Further, he obtained an LL.M. in International Commercial Law in 2013 from the University of Kent and then an LL.M. (Adv.) in 2014 in Air and Space Law from the International Institute for Air and Space Law at Leiden University. Finally. Dr Scott holds a Ph.D. from the Institute of Air Law, Space Law and Cyber Law at the University of Cologne.

Benjamyn has a strong research interest in developing technologies, having worked in this area for several years, and written numerous journal articles and book chapters on cybersecurity, unmanned aircraft, suborbital transportation and roadable aircraft. The most notable publications being his 2016 book *The Law of Unmanned Aircraft Systems: An Introduction into the Current and Future Regulation under National, Regional and International Law* and his 2019 text book *Fundamentals of International Aviation Law and Policy*.

Disclaimer

The views expressed in this work are solely that of the author in his private capacity. All those that have assisted with this work have done so in their private capacity and their views do not necessarily reflect that of any association with which they are affiliated. The content of this work is for general information purposes only and does not in any way constitute legal advice. While all attempts were made to make the information accurate at the time of writing, errors may persist and relevant information may be absent.

List of Tables and Figure

Index

D

E

Essential Air and Space Law (Series Editor: Marietta Benkö)

Volume 1: Natalino Ronzitti & Gabriella Venturini (eds.), The Law of Air Warfare – Contemporary Issues, ISBN 978-90-77596-14-2

Volume 2: Marietta Benkö & Kai-Uwe Schrogl (eds.), Space Law: Current Problems and Perspectives for Future Regulations, ISBN 978-90-77596-11-1

Volume 3: Tare Brisibe, Aeronautical Public Correspondence by Satellite, ISBN 978-90-77596-10-4

Volume 4: Michael Milde, International Air Law and ICAO, ISBN 978-90-77596-54-8

Volume 5: Markus Geisler & Marius Boewe, The German Civil Aviation Act, ISBN 978-90-77596-72-2

Volume 6: Ulrich Steppler & Angela Klingmüller, EU Emissions Trading Scheme and Aviation, ISBN 978-90-77596-79-1

Volume 7: Heiko van Schyndel (ed.), Aviation Code of the Russian Federation, ISBN 978-90-77596-80-7

Volume 8: Zang Hongliang & Meng Qingfen, Civil Aviation Law in the People's Republic of China, ISBN 978-90-77596-91-3

Volume 9: Ronald M. Schnitker & Dick van het Kaar, Aviation Accident and Incident Investigation. Concurrence of Technical, ISBN 978-94-90947-01-9

Volume 10: Michael Milde, International Air Law and ICAO, second edition, ISBN 978-90-90947-35-4

Volume 11: Ronald Schnitker & Dick van het Kaar, Safety Assessment of Foreign Aircraft Programme. A European Approach to Enhance Global Aviation Safety, ISBN 978-94-9094-793-4

Volume 12: Marietta Benkö & Engelbert Plescher, Space Law: Reconsidering the Definition/Delimitation Question and the Passage of Spacecraft through Foreign Airspace, ISBN 978-94-6236-076-1

Volume 13: Heiko van Schyndel (ed.), Aviation Code of the Russian Federation, second edition, ISBN 978-94-6236-433-2

Volume 14: Alejandro Piera Valdés, Greenhouse Gas Emissions from International Aviation: Legal and Policy Challenges, ISBN 978-94-6236-467-7

Volume 15: Peter Paul Fitzgerald, A Level Playing Field for "Open Skies': The Need for Consistent Aviation Regulation, ISBN 978-94-6236-625-1

Volume 16: Jae Woon Lee, Regional Liberalization in International Air Transport: Towards Northeast Asian Open Skies, ISBN 978-94-6236-688-6

Volume 17: Tanveer Ahmad, Climate Change Governance in International Civil Aviation: Toward Regulating Emissions Relevant to Climate Change and Global Warming, ISBN 978-94-6236-692-3

Volume 18: Michael Milde, International Air Law and ICAO, third edition, ISBN 978-94-6236-619-0

Volume 19: Nataliia Malysheva, Space Law and Policy in the Post-Soviet States, ISBN 978-94-6236-847-7

Volume 20: Philippe Clerc, Space Law in the European Context, ISBN 978-94-6236-797-5

Volume 21: Benjamyn I. Scott, Aviation Cybersecurity: Regulatory Approach in the European Union, ISBN 978-94-6236-961-0

Printed by Libri Plureos GmbH in Hamburg,
Germany

9 789462 369610